Advanced 2D Materials for Emerging Applications

Advanced 2D Materials for Emerging Applications

Guest Editor

Maiyong Zhu

MDPI

Basel • Beijing • Wuhan • Barcelona • Belgrade • Novi Sad • Cluj • Manchester

Guest Editor
Maiyong Zhu
School of Materials Science & Engineering
Jiangsu University
Zhenjiang
China

Editorial Office
MDPI AG
Grosspeteranlage 5
4052 Basel, Switzerland

This is a reprint of the Special Issue, published open access by the journal *Nanomaterials* (ISSN 2079-4991), freely accessible at: https://www.mdpi.com/journal/nanomaterials/special_issues/388JEZ8YI5.

For citation purposes, cite each article independently as indicated on the article page online and as indicated below:

Lastname, A.A.; Lastname, B.B. Article Title. *Journal Name* **Year**, *Volume Number*, Page Range.

ISBN 978-3-7258-6286-3 (Hbk)
ISBN 978-3-7258-6287-0 (PDF)
https://doi.org/10.3390/books978-3-7258-6287-0

Contents

About the Editor

Maiyong Zhu

Maiyong Zhu received his PhD degree from Yangzhou University (China) in 2011. In 2012, he started independent research work at school of Materials Science & Engineering, Jiangsu University (China), as an assistant professor. In 2015, he was promoted to associate professor. In 2020, he worked as a visiting professor at Kyoto University (Japan) under the support of the China Scholarship Council. Currently, the research of his group covers a wide range, including green strategies for synthesizing advanced functional materials for energy/environment related applications, and valorization and recycling of solid wastes.

nanomaterials

Editorial

Advanced 2D Materials for Emerging Applications

Maiyong Zhu

School of Materials Science & Engineering, Jiangsu University, Zhenjiang 212013, China; maiyongzhu@ujs.edu.cn

Since the discovery of graphene in 2004, two-dimensional (2D) materials have received increasing attention owing to their unique electronic, optical, mechanical, and chemical properties. To date, the family of 2D materials has been greatly enlarged by developing numerous emerging members, including graphdiyne, transition metal dichalcogenides (TMDs), graphitic carbon nitride (g-C_3N_4), hexagonal boron nitride (h-BN), black phosphorus (BP), MXenes, metallene, etc.

The Special Issue "Advanced 2D materials for emerging applications" represents a significant contribution to the field of 2D materials, showcasing 10 innovative studies covering a broad description of the various synthetic approaches and emerging applications. This collection offers a multidisciplinary perspective on the future of 2D materials.

The motivation behind this Special Issue is rooted in constructing various devices that take advantage of 2D materials which possess tunable interlayer spacings, adjustable terminal surface groups, and abundant surface areas to accommodate guest species, all while environmentally sustainable. The manuscripts featured in this Special Issue cover the general principles and applications of 2D materials in a wide range of devices, including supercapacitors, reconfigurable electronics, zinc-ion batteries, and radiofrequency devices. The materials discussed in the collection include graphene, TMS, MXene, and high-entropy alloys. Additionally, some papers focus on advanced manufacturing processes and strategies to improve the performance of 2D materials, such as constructing heterojunctions and introducing vacancies. Each contribution emphasizes the significance of 2D materials in boosting long-term stability, biocompatibility, and the need to be environmentally friendly.

We will now summarize these contributions in this editorial of this Special Issue. Pan et al. [1] provide a comprehensive review, discussing the applications of 2D-material-based reconfigurable electronics in logic operation and artificial intelligence summarizations by focusing on the working principles of 2D material devices used for reconfigurable electronics. Rahimi et al. [2] developed an automated pipeline for the prediction of antibacterial potential based on graphene–polymer composites. Hwang et al. [3] reported a radio frequency antenna device using single-layer graphene, which is capable of changing resonant frequency. Yang et al. [4] developed a hybrid consisting of meso-Cu-MOF and GO-COOH, in which GO-COOH serves as a substrate to support meso-Cu-MOF. When evaluated as a supercapacitor electrode, the resulting meso-Cu-MOF@GO-COOH hybrid delivers a higher capacitance of 292.5 F g^{-1}. Moreover, an asymmetric supercapacitor device is further assembled in order to verify the potential for practical application, which offers a capacitance of 63 F g^{-1} (0.5 A g^{-1}), an energy density of 27.7 Wh kg^{-1}, and a power density of 496.8 W kg^{-1}. Furthermore, the same group [5] also extended the principle to $Ti_3C_2T_x$, another important 2D material. In order to improve the structural stability of the hybrid, 2,6-diaminopyridine (DAP) and urea pyrimidinone isocyanate (UPy-NCO) units are introduced to modify the structure and properties. An electrochemical investigation indicates that the Cu-MOF@$Ti_3C_2T_X$-20%DAP-UPy hybrid exhibited excellent performance

 https://doi.org/10.3390/nano15231823

in terms of specific capacitance (148 F g^{-1} at 1 A g^{-1}), capacitance retention (88% as the current density increased from 0.2 to 5 A g^{-1}), and cycling stability (91.1% after 5000 cycles at 1 A g^{-1}).

Another important 2D material discussed in this Special Issue is metallic sulfide. Currently, much attention is focused on aqueous zinc-ion batteries due to the safety issues of lithium-ion batteries. Xu et al. [6] developed an oxygen-assisted method to synthesize phosphorus (P)-atom-embedded, three-dimensional marigold-shaped 1T MoS_2 structures (P-MoS_2), demonstrating excellent performance in a rechargeable zinc-ionic battery. The resulting P-MoS_2 possesses S vacancies (Sv). Owing to the embedment of P, the interlayer spacing of P-MoS_2 is expanded, leading to the strengthening of Zn^{2+} intercalation/deintercalation. Furthermore, the three-dimensional marigold-shaped structure with 1T phase retains an internal free space, can adapt to the volume change during charge and discharge, and improves the overall conductivity. Owing to the unique layer structures, 2D materials show great potential in the fabrication of van der Waals (vdW) heterojunctions, leading to many fascinating properties. Li et al. [7] investigated group-III selenide van der Waals (vdW) heterojunctions consisting of 2D α-In_2Se_3 and α-Ga_2Se_3 ferroelectric (FE) semiconductors employing first-principles calculations, including structural stability, electrostatic potential, interfacial charge transfer, and electronic band structures. Li et al. [8] fabricated a saturable absorber (SA) for a bulk Er:SrF_2 laser based on a few-layer SnS_2. When the average output power was 140 mW, the passively Q-switched laser achieved the shortest pulse width at 480 ns, an optimal single pulse energy at 3.78 μJ, and the highest peak power at 7.88 W. The results of the passively Q-switched laser revealed that few-layer SnS_2 had an admirable non-linear optical response, nearing a 3 μm mid-infrared solid-state laser.

For any application, the fabrication of materials is critical. For 2D materials, the fabrication at a wafer-level scale is always sought after, which allows reliable and reproducible fabrication of a large volume of devices with predictable properties. Taking this point in mind, Silva et al. [9] presented the fabrication steps for a process that allows the on-wafer fabrication of active and passive radiofrequency (RF) devices enabled by graphene. In their work, two fabrication processes are involved. In the first one, graphene is transferred to a back gate surface using critical point drying to prevent cracks in the graphene. In the second process, graphene is transferred to a flat surface planarized by ion milling, with the gate being buried beneath the graphene. In recent years, high-entropy alloys have attained much attention due to their unique properties. Traditional synthesis processes suffer from several drawbacks. Firstly, samples must be recast and pressed several times under vacuum to achieve homogeneity. Secondly, even after these processes, the elements do not mix with complete miscibility and form amalgamations in certain microscopic regions. In order to improve the synthesis efficiency and the quality of high-entropy alloys, Usman et al. [10] developed a high-vacuum radiofrequency magnetron (HVRF) sputtering process to synthesize a $Co_{30}Cr_{20}Ni_{20}Mo_{20}Ti_{10}$ high-entropy alloy.

This Special Issue is intended to serve as a valuable resource for a wide audience of researchers, engineers, industry professionals, and students who are engaged in 2D materials. I am confident that the readers will enjoy these contributions and may be able to find inspiration for their own research within this Special Issue. This series of manuscripts will provide maximum impact and will allow researchers in other areas to apply the same methodologies in order to understand the mechanisms of self-assembly in their systems.

I am grateful to all the authors for submitting their studies to the present Special Issue and for its successful completion. I also thank the *Nanomaterials* reviewers for enhancing the quality and impact of all submitted papers. Finally, I sincerely thank the editorial staff of *Nanomaterials* for their support during the development and publication of this Special Issue.

Conflicts of Interest: The author declares no conflict of interest.

References

1. Pan, J.; Zhang, Y.; Yin, J.; Guo, P.; Yang, Y.; Ren, T.L. Principles and Applications of Two-Dimensional Semiconductor Material Devices for Reconfigurable Electronics. *Nanomaterials* **2025**, *15*, 201. [CrossRef] [PubMed]
2. Rahimi, S.; Lovmar, T.; Aulova, A.; Pandit, S.; Lovmar, M.; Forsberg, S.; Svensson, M.; Kádár, R.; Mijakovic, I. Automated Prediction of Bacterial Exclusion Areas on SEM Images of Graphene–Polymer Composites. *Nanomaterials* **2023**, *13*, 1605. [CrossRef] [PubMed]
3. Hwang, H.J.; Kim, S.-Y.; Lee, S.K.; Lee, B.H. Reconfigurable Single-Layer Graphene Radio Frequency Antenna Device Capable of Changing Resonant Frequency. *Nanomaterials* **2023**, *13*, 1203. [CrossRef] [PubMed]
4. Yang, M.; Zhang, Y.; Li, W.; Ye, P.; Nie, Y.; Zhu, M.; Li, S. Hierarchical 2D Cu-MOF@Graphene-Based Hybrids for Supercapacitor Electrodes. *Nanomaterials* **2025**, *15*, 1628. [CrossRef] [PubMed]
5. Li, S.; Qu, X.; Liu, F.; Ye, P.; Yang, B.; Cheng, Q.; Yang, M.; Nie, Y.; Zhu, M. Optimizing the Structure and Performances of Cu-MOF@$Ti_3C_2T_x$ Hybrid Electrodes by Introducing Modulated Ligand. *Nanomaterials* **2025**, *15*, 864. [CrossRef] [PubMed]
6. Xu, Q.; Li, X.; Wu, L.; Zhang, Z.; Chen, Y.; Liu, L.; Cheng, Y. Enlarged Interlayer Spacing of Marigold-Shaped 1T-MoS_2 with Sulfur Vacancies via Oxygen-Assisted Phosphorus Embedding for Rechargeable Zinc-Ion Batteries. *Nanomaterials* **2023**, *13*, 1185. [CrossRef] [PubMed]
7. Li, P.; Kong, D.; Yang, J.; Cui, S.; Chen, Q.; Liu, Y.; He, Z.; Liu, F.; Xu, Y.; Wei, H.; et al. Engineering Nonvolatile Polarization in 2D α-In_2Se_3/α-Ga_2Se_3 Ferroelectric Junctions. *Nanomaterials* **2025**, *15*, 163. [CrossRef] [PubMed]
8. Li, C.; Yang, Q.; Zu, Y.; Din, S.Z.U.; Yue, Y.; Zhai, R.; Jia, Z. SnS_2 as a Saturable Absorber for Mid-Infrared Q-Switched Er:SrF_2 Laser. *Nanomaterials* **2023**, *13*, 1989. [CrossRef] [PubMed]
9. Silva, V.; Colmiais, I.; Dinis, H.; Borme, J.; Alpuim, P.; Mendes, P.M. Wafer-Level Fabrication of Radiofrequency Devices Featuring 2D Materials Integration. *Nanomaterials* **2025**, *15*, 1119. [CrossRef] [PubMed]
10. Usman, K.; Kang, D.; Jeong, G.; Alam, K.; Raveendran, A.; Ser, J.; Jang, W.; Cho, H. The Surface Properties of Implant Materials by Deposition of High-Entropy Alloys (HEAs). *Nanomaterials* **2023**, *13*, 1123. [CrossRef] [PubMed]

nanomaterials

MDPI

Article

Hierarchical 2D Cu-MOF@Graphene-Based Hybrids for Supercapacitor Electrodes

Mengkun Yang [1], Yongqiang Zhang [1], Wenjie Li [1], Pingwei Ye [2,*], Yijing Nie [1], Maiyong Zhu [1] and Sumin Li [1,*]

[1] School of Materials Science & Engineering, Jiangsu University, Zhenjiang 212013, China; 2222305060@stmail.ujs.edu.cn (M.Y.); 2212005078@stmail.ujs.edu.cn (Y.Z.); 2222305089@stmail.ujs.edu.cn (W.L.); nieyijing@ujs.edu.cn (Y.N.); maiyongzhu@ujs.edu.cn (M.Z.)

[2] State Key Laboratory of NBC Protection for Civilian, Beijing 102205, China

* Correspondence: yepw2001@163.com (P.Y.); li_sm@ujs.edu.cn (S.L.)

Abstract

Recently, two-dimensional metal–organic framework (2D MOF) hybrids are attracting much attention in supercapacitors. However, their performance is limited by the insufficient utilization of active sites and poor conductivity. Notably, the full utilization of active sites highly depends on the fast transport and diffusion of ions. Here, a Cu-MOF@GO-COOH hybrid was constructed, with GO-COOH as the substrate, to promote electron transfer, and Meso-Cu-MOF@GO-COOH was further obtained by introducing mesopores inside nanosheets to optimize the transportation paths for ions. The GO-COOH substrate improves the capacitance by enhancing the surface capacitive behavior, while the mesopores improve the charge-storage capacity by enhancing the diffusive behavior. The as-obtained Meso-Cu-MOF@GO-COOH exhibits a higher capacitance of 292.5 F g^{-1} compared with Cu-MOF@GO-COOH (193.7 F g^{-1}) and 2D Cu-MOF (141.4 F g^{-1}) at a current density of 1 A g^{-1}. Moreover, the prepared Cu-MOF@GO-COOH//AC device delivers a capacitance of 63 F g^{-1} (0.5 A g^{-1}), an energy density of 27.7 Wh kg^{-1}, and a power density of 496.8 W kg^{-1}, showing a great potential for practical applications.

Keywords: Cu-MOF; GO-COOH; soft-template; supercapacitor

1. Introduction

Benefiting from their high-power density, rapid charge–discharge process, greenness, and long lifetime, supercapacitors are gaining great attention as next-generation energy storage devices. To enhance their electrochemical performance, various materials, including traditional materials (carbon-based, conducting polymers, and metal oxides) and novel nanomaterials (quantum dots, transition metal dichalcogenides, Mxenes, etc.), have been developed as supercapacitor electrodes [1,2]. Metal–organic frameworks (MOFs) are a novel porous material, formed via the coordination reaction between metal ions and organic ligands. Benefiting from their diverse pore architecture, high surface area, and tunable functionality, MOFs are being intensively investigated in energy storage [3–8], detection [9–13], sensing [14–18], and adsorption [19], etc. Notably, owing to the highly accessible active sites, two-dimensional MOFs (2D MOFs), which are prone to producing capacitance, are drawing much attention in the field of supercapacitor electrodes [20–25].

However, the electrochemical performance of 2D MOFs is restricted by their unsatisfactory conductivity, stacking problems, and insufficient utilization of active sites, limiting their application as energy-storing electrodes. To overcome these shortcomings, some conductive materials, such as graphene-based and $Ti_3C_2T_x$ substrates, are preferred to construct MOF-based hybrids [26–28]. By combining the merits of 2D MOFs with those of conductive materials, the electrochemical performances of MOF hybrids have been certainly enhanced [29–31].

To further shorten the gap between the practical and theoretical performance of 2D MOFs, maximizing the utilization of active sites is imperative. Notably, the hierarchically porous architecture has been proved to be an effective strategy to enhance electrode performance by promoting ion transport and diffusion [32–36]. By promoting the accessibility of ions to micropores and small-sized mesopores, the utilization of active sites is improved.

In this paper, to optimize the transport paths and to promote the movement of ions directly through 2D MOF nanosheets rather than around them, a soft template was used to produce mesopores inside MOF nanosheets, thus providing convenient channels for ion transport. Moreover, as a low-cost and environmentally friendly material, Cu-based MOFs present great potential in developing high-performance electrodes [37].

As an amphiphilic block copolymer, polystyrene-b-poly(ethylene oxide) presents a prominent merit, in that it has adjusting pore sizes, in which changing lengths of hydrophobic chain segments lead to various pores [38,39]. Here, a polystyrene-b-poly(ethylene oxide) (PS_{102}-b-PEO_{114}) was chosen as the soft template to produce mesopores inside MOF nanosheets. Benefiting from the ample carboxyl groups on the surface of GO-COOH, hydrogen bonds can be formed between GO-COOH and the micelles to promote micelles to arrange on the GO-COOH surface and to further guide the growth of Cu-MOF. After the micelles were removed, mesopores were left inside MOF nanosheets to obtain hierarchical Meso-Cu-MOFs@GO-COOH, in which new transport channels are provided for ions and their diffusion ability is enhanced, improving the effective utilization of active sites inside MOFs.

2. Experimental Methods

2.1. Synthesis of Materials

The synthetic methods of 2D Cu-MOF, GO-COOH, and BCP@GO-COOH are described in the Supplementary Materials Section.

Synthesis of Cu-MOF@GO-COOH. Firstly, 20 mg of $Cu(OAc)_2 \bullet H_2O$ (>98%) was dissolved in a mixed solvent containing 2 mL of N, N-dimethylformamide (DMF, 99.5%), and 4 mL of ethanol absolute (EtOH, 99.7%). Next, 20 mg of H_2BDC-NH_2 (98.6%) and 10 mg of GO-COOH were added to another solvent containing 10 mL of DMF and 5 mL of EtOH to form a mixture. The copper–salt solution was slowly added to the mixture and stirred continuously for 30 min. After that, the product was collected, washed with DMF and EtOH, and dried at 60 °C for 24 h, to obtain Cu-MOF@GO-COOH.

Synthesis of Meso-Cu-MOF@GO-COOH. Firstly, 20 mg of $Cu(OAc)_2 \bullet H_2O$ was dissolved in a mixture containing 2 mL of DMF and 4 mL of EtOH. Next, 20 mg of H_2BDC-NH_2 was added to the BCP@GO-COOH solution and stirred continuously for 60 min to form a mixture. Subsequently, the copper–salt solution was slowly added to the mixture and stirred continuously for 30 min. After that, the product was collected, and then washed with DMF, tetrahydrofuran (THF, 99.5%), and EtOH. At last, the as-obtained product was dried at 60 °C for 24 h to obtain Meso-Cu-MOF@GO-COOH (Figure 1).

Notably, the Cu-MOF@BCP@GO-GOOH can be obtained by washing with just DMF and EtOH, without THF.

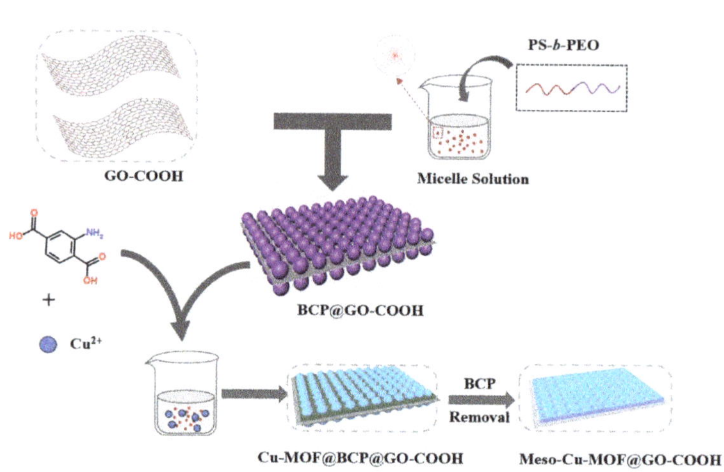

Figure 1. Schematic illustration of preparing Meso-Cu-MOF@GO-COOH.

2.2. Structural Analysis

Scanning electron microscopy (SEM, JEOL JXA-840A) and transmission electron microscopy (TEM, JEOL JEM-2100 PLUS) were used to analyze the surface morphology of samples. Fourier transforms infrared (FTIR) and X-ray diffraction (XRD) were performed to investigate the elemental composition of materials. The specific surface area and pore characteristic of samples were analyzed via a physisorption analyzer (ASAP 2020M).

2.3. Electrochemical Characterization

The electrochemical performances of samples were tested on an electrochemical workstation (CHI760E), including cyclic voltammetry (CV), galvanostatic charge–discharge (GCD), and electrochemical impedance spectroscopy (EIS). Using a three-electrode system to estimate the performance of single electrode, in which a working electrode (the as-prepared materials), counter electrode (Pt plate), and reference electrode (Ag/AgCl) were used. Additionally, 3 M KOH aqueous solution was used as the electrolyte.

The specific capacitance of electrode was calculated based on the following formula [40]:

$$C\,(\mathrm{F\,g^{-1}}) = I \int (1/m \times V(t))\,\mathrm{d}t = I\,\Delta t/m\,\Delta V \tag{1}$$

in which I, Δt, and ΔV correspond to the current, discharge time, and voltage, respectively, while m is the mass of active materials.

An asymmetric supercapacitor (ASC) was prepared using Meso-Cu-MOF@GO-COOH and activated carbon (AC) as electrodes. The preparation process of working electrodes is displayed in the Supplementary Materials. The loading mass of the active material was 1–1.5 mg cm^{-2}. To maximize the performance of device, the optimal mass ratio of two electrodes was calculated according to the following formula:

$$\frac{m_+}{m_-} = \frac{C_- \times \Delta V_-}{C_+ \times \Delta V_+} \tag{2}$$

in which m, C, and ΔV correspond to the mass, specific capacitance, and voltage of the positive and negative electrodes, respectively. Moreover, the capacitance value of the as-assembled ASC device can also by estimated according to Equation (1), except, in that case, m would refer to the mass of two electrodes. The energy density (E) and power density (P) of the ASC device are calculated via the following formulas [41]:

$$E = \frac{C \times \Delta V^2}{2 \times 3.6} \quad (3)$$

$$P = \frac{3600 \times E}{\Delta t} \quad (4)$$

in which E is the energy density, P is the power density, Δt is the discharge time, and ΔV is the potential window.

3. Results and Discussion

3.1. Structural Characteristics

The microstructure of GO-COOH is shown in Figure 2a; many wrinkles can be noticed. Furthermore, when using GO-COOH as the substrate, promoted by the hydrogen bonding force, the soft template (PS_{102}-b-PEO_{114}) self-assembled into uniform spherical micelles (Figure 2b), and covered the surface of GO-COOH substrate. Notably, when the length of the hydrophobic chain segment in PS_n-b-PEO_{114} (n = 70, 90, 102 and 150) was changed, micelles with different sizes were obtained (Figure S1), by which the sizes of mesopores can be regulated. Figure 2c displays the morphology of pristine 2D Cu-MOF, in which uniform nanosheets with a size of 100~200 nm can be seen. For the Cu-MOF@GO-COOH sample, the size of the nanosheets (Figure 2d) is somewhat larger than that of the pristine 2D Cu-MOF, presenting an average size of about 300 nm, which may be related to the induced growth originating from the functional groups on the GO-COOH surface (Figure S2). When using BCP@GO-COOH as the substrate to prepare 2D Cu-MOF, large numbers of spherical micelles could still be identified (Figure 2e). During the preparation of MOF nanosheets, the coordination force could be produced between the hydrophilic segments of micelles and Cu^{2+}, forming the nuclear and further guiding the growth of MOF nanosheets, and, simultaneously, wrapping the micelles. THF was used to remove the BCP micelles, leaving abundant pores inside the MOF nanosheets (Figure 2f).

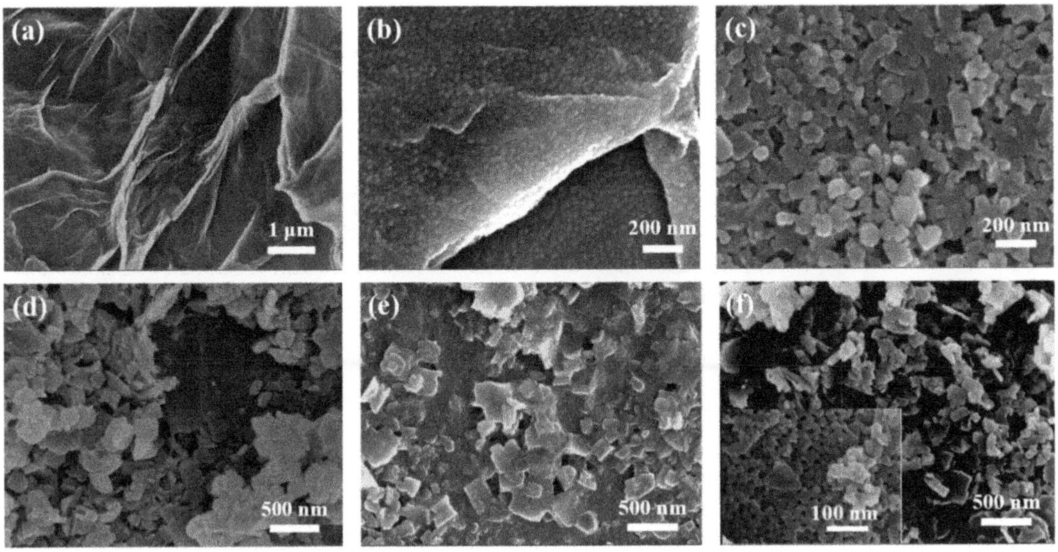

Figure 2. SEM images of samples: (**a**) GO-COOH, (**b**) BCP@GO-COOH, (**c**) 2D Cu-MOF, (**d**) Cu-MOF@GO-COOH, (**e**) Cu-MOF@BCP@GO-COOH, and (**f**) Meso-Cu-MOF@GO-COOH (the insert is high-resolution SEM image).

To further understand the crystal structure of different samples, XRD analysis was performed. As shown in Figure 3, the peaks at 10.3, 11.8, 16.7, 20.8, and 24.8° correspond to the (001), (100), (120), (121), and (131) crystal planes, respectively, which is basically consistent with those of the simulated MOF-46 crystals, proving the success synthesis of 2D Cu-MOF [42,43]. After combining 2D Cu-MOF with GO-COOH, though the corresponding peak intensity decreased slightly, the peak position of Cu-MOF@GO-COOH showed no obvious difference with that of the 2D Cu-MOF. Similarly, compared with Cu-MOF@GO-COOH, Meso-Cu-MOF@GO-COOH presented a very similar peak position except for a slight decrease in peak intensity, which suggests that the crystal structure of Cu-MOF was not obviously changed during the introduction of the mesoporous structure using a soft template.

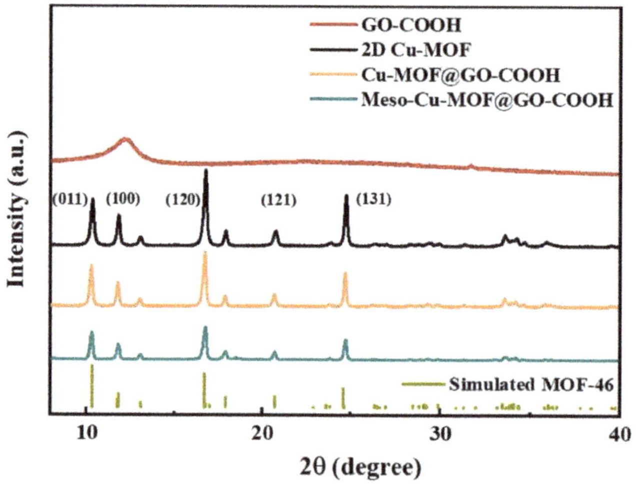

Figure 3. XRD patterns of samples.

The pore characteristics of samples are displayed in Figure 4. The materials exhibit type I isotherms as well as obvious hysteresis loops, indicating the architecture of hierarchical pores. The type I isotherm shape in low pressure suggests the presence of micropores, while the obvious H_3-type hysteresis loop in higher pressure suggests the presence of large-sized slit pores caused by the accumulation of 2D Cu-MOF nanosheets. As displayed in Figure 4b, similar pore distribution, including micropore (1.2 nm), mesopores (18 and 34 nm), as well as macropores (51 and 69 nm), can be noticed. The proportion of the micropores (1.2 nm) decreased greatly while a new mesopore (14 nm) was formed in Meso-Cu-MOF@GO-COOH. Upon the removal of the BCP micelles, a new mesopore was produced inside the 2D Cu-MOF nanosheets (Figure 2f), which could provide interpenetrated channels for ion transport and subsequently make it easier to access the active sites in MOF.

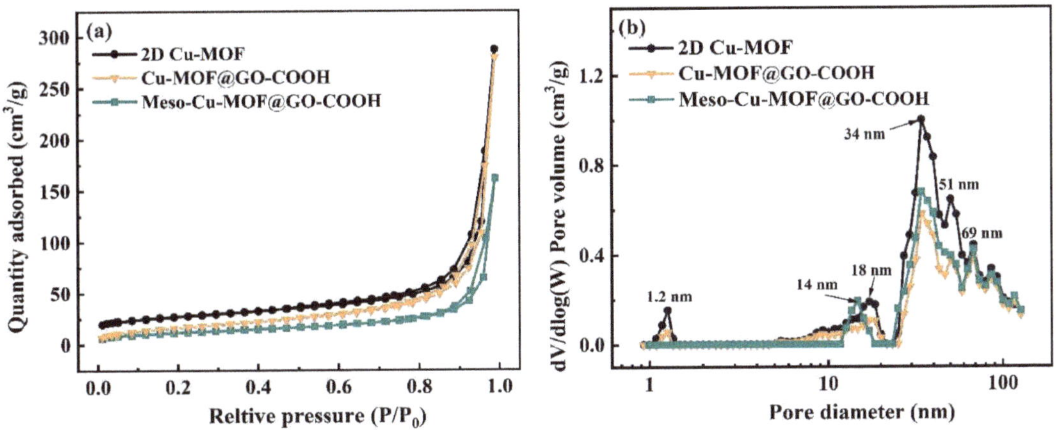

Figure 4. (**a**) N_2 adsorption–desorption isotherms and (**b**) corresponding BJH pore distribution plots of samples.

The infrared spectra of samples are shown in Figure 5. Obviously, for GO-COOH, the broad peak at 3666~3245 cm^{-1}, corresponding to the stretching vibration of -OH, was strengthened compared with GO, which is associated with the introduction of -COOH [44–47]. For 2D Cu-MOF and its hybrids, the two peaks at 3364 and 3486 cm^{-1} correspond to the symmetric and asymmetric stretching vibrations of the N-H groups in H_2BDC-NH_2, respectively. The peak of C=O in 2D Cu-MOF presented a blue shift to 1669 cm^{-1} compared with that in GO and GO-COOH, which is related to the coordination with Cu^{2+} ions, also proving the success synthesis of 2D Cu-MOF. Because the characteristic functional groups of PS_{102}-b-PEO_{114} are similar to those of H_2BDC-NH_2, the characteristic peaks of Cu-MOF@BCP@GO-COOH presented no obvious difference to those of Cu-MOF@GO-COOH and Meso-Cu-MOF@GO-COOH. However, in the above-mentioned SEM analysis (Figure 2), the difference in the microstructure of the as-prepared samples has been clearly displayed.

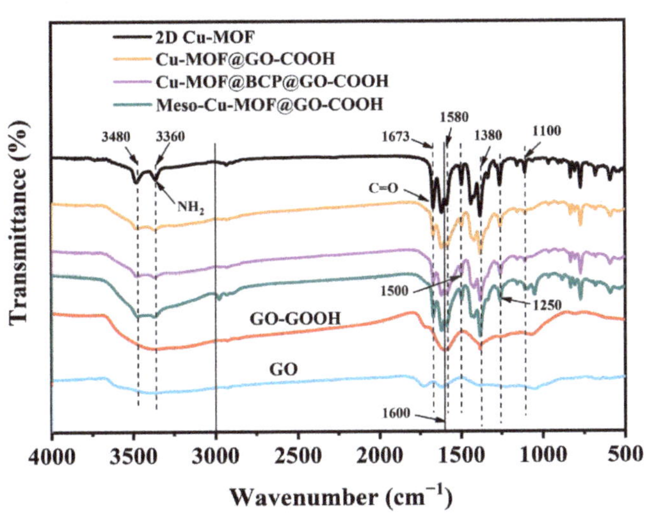

Figure 5. FTIR spectra of samples.

3.2. Electrochemical Performance of 2D MOF Hybrid Electrodes

The electrochemical performances of the prepared materials are shown in Figure 6. According to the CV and GCD curves, the energy storage process of the hybrids consists of pseudocapacitive and double-layer behaviors. The enclosed region in CV curves (Figure 6a) reveals the ability of the samples to store charges. Furthermore, at a current density of 1 A g^{-1} (Figure 6b), the capacitance values of 2D Cu-MOF, Cu-MOF@GO-COOH, and Meso-Cu-MOF@GO-COOH were calculated to be 141.4, 193.7, and 292.5 F g^{-1} (1 A g^{-1}), respectively, which can be explained by the Nyquist plots and equivalent circuit in Figure 6c. Clearly, the two hybrids exhibit a smaller intercept at the real axis compared with 2D Cu-MOF, implying lower internal resistance (*R*s) owed to the introduction of GO-COOH substrate. The pi-pi interaction between 2D Cu-MOF and GO-COOH nanosheets could enhance carrier transport, which is conducive to improving conductivity. Moreover, in the low-frequency region, Meso-Cu-MOF@GO-COOH presents a steeper slope of the straight line than those of the other samples, suggesting faster ion diffusion rates [48–50], which is attributed to the convenient channels for ion transport stemming from the interpenetrated holes inside MOF nanosheets. For the equivalent circuit, R_s means the electrolyte resistance, C_d presents the double-layer capacitance, R_{ct} is the polarization resistance stemmed from charge transfer, and Z_w is the Warburg resistance associated with ion diffusion. For Meso-Cu-MOF@GO-COOH, with the increased scan rates, the shapes of the CV curves present no obvious change (Figure 6d), suggesting fast electron transport inside the active material [51]. According to the GCD curves (Figure 6e), the specific capacitance values of Meso-Cu-MOF@GO-COOH were calculated to be 307.3 and 286.1 F g^{-1} as the current density increased from 0.2 to 5 A g^{-1}, presenting a 93% capacitance retention (Figure 6f). The as-obtained Meso-Cu-MOF@GO-COOH presents enhanced performance compared with some reported Cu-MOF electrodes ($Cu_3(HHTP)_2$, 110–114 F g^{-1} at 0.04–0.05 A g^{-1} [52]; Cu@BTC, 228 F g^{-1} at 1.5 A g^{-1} [53]; and Cu-MOF, 37.91 F g^{-1} at 0.8 A g^{-1} [54], etc.).

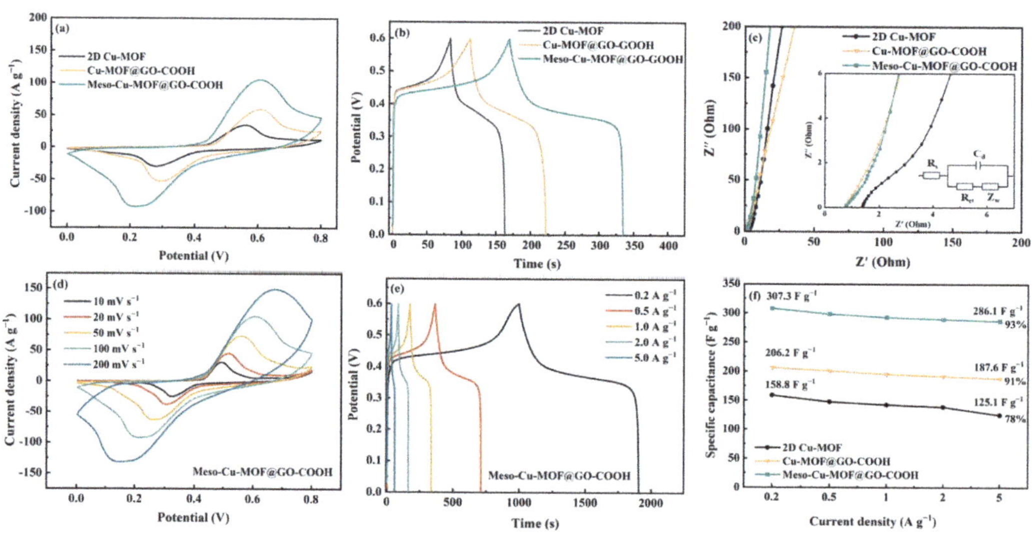

Figure 6. Electrochemical performances of samples: (**a**) CV curves at a scan rate of 100 mV s^{-1}, (**b**) GCD curves at a current density of 1 A g^{-1}, and (**c**) Nyquist plots of samples. (**d**) CV curves at various scan rates, (**e**) GCD curves at various current densities, and (**f**) the specific capacitance as a function of current density of samples.

To further investigate the cycling stability of the Meso-Cu-MOF@GO-COOH hybrid, 2000 cycles were performed at a current density of 2 A g^{-1} and the results are shown in Figure S3, delivering an 82% capacitance retention.

3.3. Energy Storage Mechanism

By integrating GO-COOH into 2D MOF to construct 2D MOF hybrids and subsequently producing pores inside MOF nanosheets, Cu-MOF@GO-COOH and Meso-Cu-MOF@GO-COOH were endowed with enhanced capacitive properties. To further investigate the energy storage mechanism, the *b*-value model [55] and Dunn method [56] were used to reveal the kinetic behaviors. The peak current (i) and scan rate (v) can be described in the following formula in the *b*-value model:

$$i = a \cdot v^b \quad (5)$$

Here, b is a variable with a range of 0.5~1.0, and the 0.5 value corresponds to the diffusive-controlled behavior, while the 1.0 corresponds to the surface capacitive behavior.

As shown in Figure 7a,d, for Cu-MOF@GO-COOH, the *b*-values of both oxidation and the reduction peaks are close to those of 2D Cu-MOF, implying that the two samples present similar charge-storage behaviors. Differently, the *b*-values of Meso-Cu-MOF@GO-COOH are closer to the critical value of 0.5, indicating that the diffusive-controlled behavior is enhanced, which can be confirmed by the Dunn method's results. The contributions of the two charge-storage behaviors can be identified via the Dunn model [57,58]:

$$i(V) = k_1 v + k_2 v^{0.5} \quad (6)$$

where $k_1 v$ and $k_1 v^{0.5}$ correspond to the surface capacitive current and diffusive current, respectively, by which the charge-storage process of electrode materials can be investigated. Clearly, at a scan rate of 10 mV s^{-1} (Figure 7b,e,h), the surface capacitive ratios of 2D Cu-MOF, Cu-MOF@GO-COOH, and Meso-Cu-MOF@GO-COOH are 44%, 46%, and 36%, respectively. For 2D Cu-MOF and Cu-MOF@GO-COOH, the surface capacitive percentages increased greatly with the increased scan rates, obtaining 77% and 86%, respectively, at a scan rate of 200 mV s^{-1}. Utilizing GO-COOH to construct 2D MOF hybrids accelerates the charge transfer, thus lowering the internal resistance (Figure 6c), and thereby enhancing the surface capacitive behavior, which may be the main reason for the improved capacitance value of Cu-MOF@GO-COOH. Notably, for Meso-Cu-MOF@GO-COOH, even when the scan rate reached 50 mV s^{-1}, the ratio of diffusive current was still higher than that of surface capacitive current, implying the dominant diffusion-controlled behavior. Owing to the interpenetrated pores, ions could diffuse adequately inside Meso-Cu-MOF@GO-COOH, allowing the active sites to be utilized more fully and conducting sufficient redox reactions inside MOF nanosheets. Thus, benefiting from the enhanced diffusion behavior, the electrochemical performance of Meso-Cu-MOF@GO-COOH was further improved.

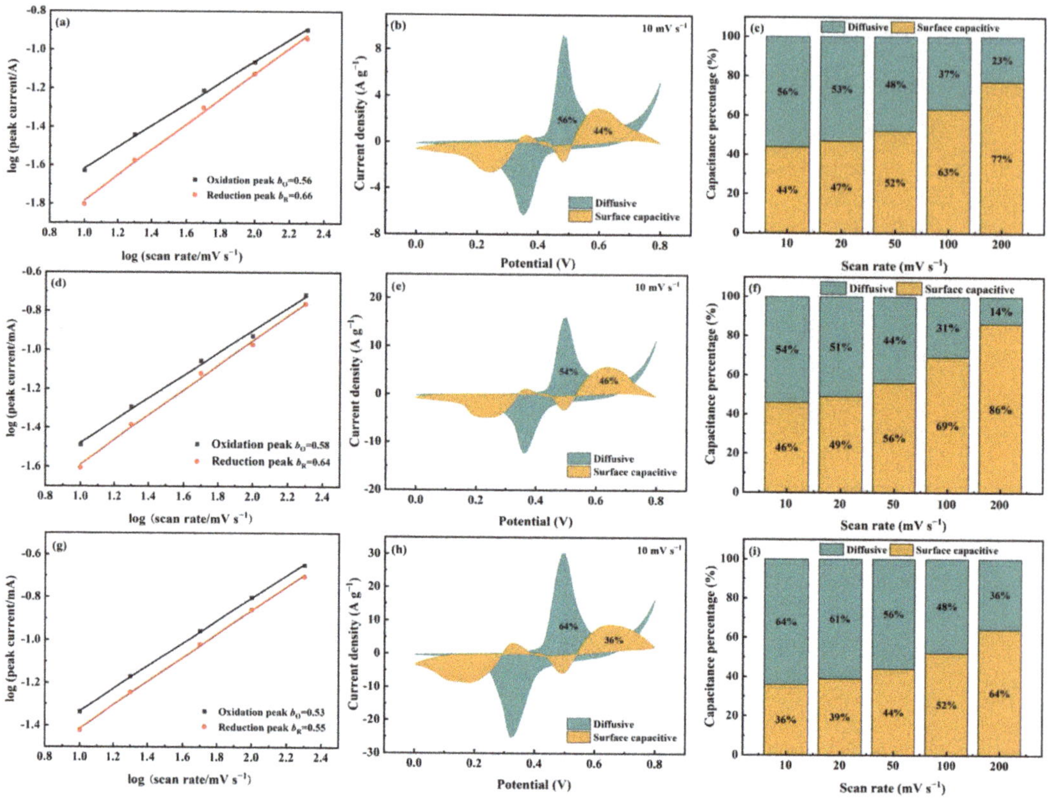

Figure 7. The b-value and Dunn model analysis of (**a–c**) 2D Cu-MOF, (**d–f**) Cu-MOF@GO-COOH, and (**g–i**) Meso-Cu-MOF@GO-COOH. (**a**,**d**,**g**) Plots of log(i) against log(v), (**b**,**e**,**h**) surface capacitive and diffusive contributions at a scan rate of 10 mV s^{-1}, and (**c**,**f**,**i**) capacitance contribution ratios at different scan rates.

3.4. Electrochemical Performance of ASC Device

An ASC device was prepared using Meso-Cu-MOF@GO-COOH and activated carbon (AC) as the positive and negative electrodes, respectively. Combining the capacitance values of AC, calculated via the GCD curves, with the potential ranges of two electrodes (Figure S4), the optimal mass ratio (m_+/m_-) of 1.2 was chosen to prepare the ASC device. Furthermore, CV tests under various voltage windows showed that the curve shape presented no deformation even at 1.8 V (Figure 8a), exhibiting a broad working voltage window. With the scan rate was increased to 100 mV s^{-1} (Figure S4), no obvious changes can be noticed, indicating fast charge transfer and ion transport.

The capacitance values corresponding to 0.5, 1, 2, 3, and 5 A g^{-1} were calculated to be 63, 60.9, 57.2, 54.3, and 50.9 F g^{-1}, respectively, and the 81% capacitance retention rate shows a relatively good performance rate (Figure 8b,c). The device delivers a maximum energy density of 27.7 Wh kg^{-1} at a power density of 496.8 W kg^{-1}, which are figures comparable to or surpassing many of the recently reported MOF devices mentioned in Figure 8d [51,59–62]. Moreover, the cycle performance was investigated at a current density of 3 A g^{-1}, delivering a capacitance retention of 84.2% after 2000 cycles (Figure 8e,f).

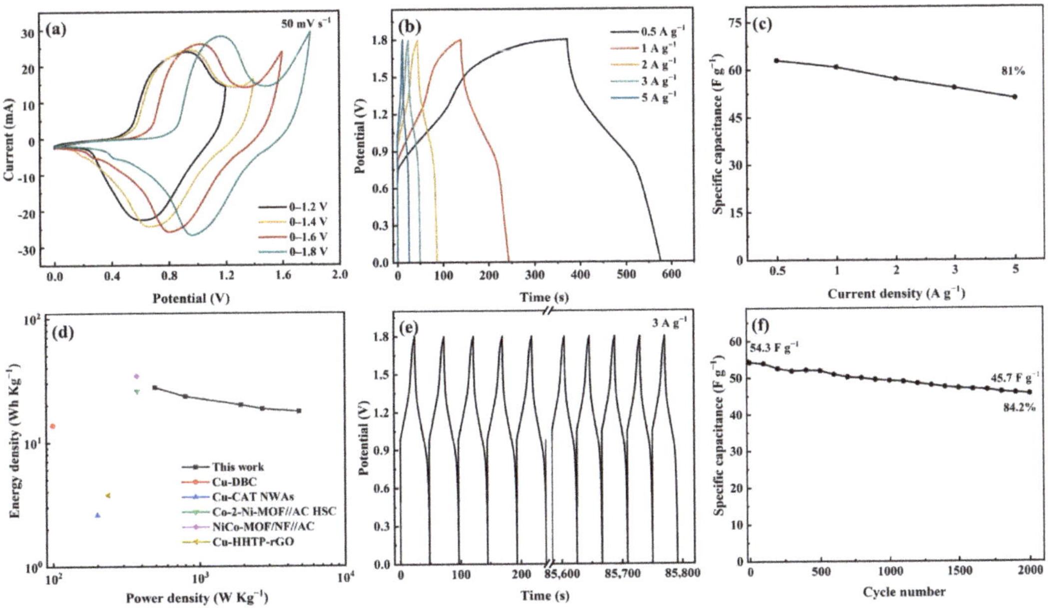

Figure 8. Electrochemical performances of the ASC device. (**a**) CV curves at different potential windows, (**b**) GCD curves at different current densities, (**c**) specific capacitance under various current densities, (**d**) Ragone plots, (**e**) cycling GCD curves, and (**f**) cycling performance.

4. Conclusions

Here, a hierarchical Meso-Cu-MOFs@GO-COOH was synthesized and used as supercapacitor electrode. The introduction of GO-COOH accelerates electron transfer, lowers the internal resistance of the hybrid, and thus enhances the charge storage capacity by promoting the surface capacitive behavior. Additionally, the interpenetrated mesopores inside MOF sheets optimize the diffusion paths of ions, improving the energy storage capacitance via enhancing the diffusive behavior. The Meso-Cu-MOFs@GO-COOH exhibited an increased capacitance of 292.5 F g^{-1} (1 A g^{-1}), which is 1.5 and 2.1 times as high as that of Cu-MOFs@GO-COOH and 2D Cu-MOF, respectively. When the current density increased from 0.2 to 5 A g^{-1}, an excellent capacitance retention of 93% was obtained. Furthermore, the assembled Cu-MOF@GO-COOH//AC supercapacitor exhibited a capacitance of 63 F g^{-1} (0.5 A g^{-1}), an energy density of 27.7 Wh kg^{-1}, and a power density of 496.8 W kg^{-1}, showing its practical potential in energy storage applications.

Supplementary Materials: The following supporting information can be downloaded at: https://www.mdpi.com/article/10.3390/nano15211628/s1, Figure S1: SEM of BCP@GO-COOH samples. (a) PS_{70}-BCP@GO-COOH, (b) PS_{90}-BCP@GO-COOH, (c) PS_{102}-BCP@GO-COOH and (d) PS_{150}-BCP@GO-COOH; Figure S2: TEM image of Cu-MOF@GO-COOH; Figure S3: Cycling test of Meso-Cu-MOF@GO-COOH: (a) GCD curves at a current density of 2 A g-1 and (b) cycling capacitance; Figure S4: (a) GCD curves of AC electrode, (b) CV curves of AC and Meso-Cu-MOF@GO-COOH, and (c) CV curves of Meso-Cu-MOF@GO-COOH at different scan rates.

Author Contributions: Methodology, P.Y., Y.N. and M.Z.; validation, Y.Z. and W.L.; writing—original draft preparation, M.Y.; writing—review and editing, S.L. All authors have read and agreed to the published version of the manuscript.

Funding: This research was funded by the National Natural Science Foundation of China (52073126, 52173020).

Data Availability Statement: The original contributions presented in this study are included in the article and Supplementary Material. Further inquiries can be directed to the corresponding authors.

Acknowledgments: We are thankful to the National Natural Science Foundation of China (52073126, 52173020).

Conflicts of Interest: The authors declare no conflicts of interest.

References

1. Bhaduri, A.; Kim, C.; Ha, T. Recent Developments in Materials Design for Advanced Supercapacitors. *Energy Environ. Mater.* **2025**, *8*, e70070. [CrossRef]
2. Tsyganov, A.; Vikulova, M.; Zotov, I.; Korotaev, E.; Plugin, I.; Sysoev, V.; Kirilenko, D.; Rabchinskii, M.; Asoyan, A.; Gorokhovsky, A.; et al. Application of $W_{1.33}CT_z$ MXenes obtained by hydrothermal etching as an additive to enhance the electrochemical energy storage properties of binder-free $Ti_3C_2T_x$ MXene films. *Dalton Trans.* **2025**, *54*, 8547–8558. [CrossRef]
3. Xia, W.; Mahmood, A.; Zou, R.; Xu, Q. Metal-organic frameworks and their derived nanostructures for electrochemical energy storage and conversion. *Energy Environ. Sci.* **2015**, *8*, 1837–1866. [CrossRef]
4. Qiu, T.; Liang, Z.; Guo, W.; Tabassum, H.; Gao, S.; Zou, R. Metal–Organic Framework-Based Materials for Energy Conversion and Storage. *ACS Energy Lett.* **2020**, *5*, 520–532. [CrossRef]
5. Mohamed, A.M.; Sayed, D.M.; Allam, N.K. Optimized Fabrication of Bimetallic ZnCo Metal-Organic Framework at NiCo-Layered Double Hydroxides for Multiple Storage and Capability Synergy All-Solid-State Supercapacitors. *ACS Appl. Mater. Interfaces* **2023**, *15*, 16755–16767. [CrossRef] [PubMed]
6. Yue, L.; Chen, L.; Wang, X.; Lu, D.; Zhou, W.; Shen, D.; Yang, Q.; Xiao, S.; Li, Y. Ni/Co-MOF@aminated MXene hierarchical electrodes for high-stability supercapacitors. *Chem. Eng. J.* **2023**, *451*, 138687. [CrossRef]
7. Li, S.; Wang, R.; Xie, M.; Xu, Y.; Chen, J.; Jiao, Y. Construction of trifunctional electrode material based on Pt-Coordinated Ce-Based metal organic framework. *J. Colloid Interface Sci.* **2022**, *611*, 378–389. [CrossRef]
8. Wang, D.-G.; Liang, Z.; Gao, S.; Qu, C.; Zou, R. Metal-organic framework-based materials for hybrid supercapacitor application. *Coord. Chem. Rev.* **2020**, *404*, 213093. [CrossRef]
9. Liang, N.; Shi, B.; Hu, X.; Li, W.; Huang, X.; Li, Z.; Zhang, X.; Zou, X.; Shi, J. A ternary heterostructure aptasensor based on metal-organic framework and polydopamine nanoparticles for fluorescent detection of sulfamethazine. *Food Chem.* **2024**, *460*, 140570. [CrossRef]
10. Xu, S.; Shao, D.; Wang, J.; Zheng, X.; Yang, Z.; Wang, A.; Chen, Z.; Gao, Y. Pre-ligand-induced porous MOF as a peroxidase mimic for electrochemical analysis of deoxynivalenol (DON). *Food Chem.* **2025**, *480*, 143860. [CrossRef]
11. Liang, N.; Shi, B.; Hu, X.; Shi, Y.; Wang, T.; Huang, X.; Li, Z.; Zhang, X.; Zou, X.; Shi, J. Simultaneous adsorption and fluorescent sensing of ampicillin based on a trimetallic metal-organic framework. *Food Chem.* **2025**, *472*, 142891. [CrossRef]
12. Guan, H.; Chen, Y.; Wang, D.; Liu, Q.; Zhong, J.; Zhang, Z.; Lü, D. The novel nanozyme-based electrochemical-driven electrochromic visual biosensor based on PEDOT: PSS/RGO conductive film for rapid detection of nitrite in food samples. *Food Chem.* **2025**, *481*, 143971. [CrossRef] [PubMed]
13. Zhou, R.; Wu, X.; Xue, S.; Yin, L.; Gao, S.; Zhang, Y.; Wang, C.; Wang, Y.; El-Seedi, H.R.; Zou, X.; et al. Magnetic metal-organic frameworks-based ratiometric SERS aptasensor for sensitive detection of patulin in apples. *Food Chem.* **2025**, *466*, 142200. [CrossRef]
14. Shi, B.; Zhang, X.; Li, W.; Liang, N.; Hu, X.; Xiao, J.; Wang, D.; Zou, X.; Shi, J. An intrinsic dual-emitting fluorescence sensing toward tetracycline with self-calibration model based on luminescent lanthanide-functionalized metal-organic frameworks. *Food Chem.* **2023**, *400*, 133995. [CrossRef] [PubMed]
15. Liang, N.; Hu, X.; Zhang, X.; Li, W.; Guo, Z.; Huang, X.; Li, Z.; Zhang, R.; Shen, T.; Zou, X.; et al. Ratiometric sensing for ultratrace tetracycline using electrochemically active metal–organic frameworks as response signals. *J. Agric. Food Chem.* **2023**, *71*, 7584–7592. [CrossRef]
16. Chen, X.; Xu, J.; Li, Y.; Zhang, L.; Bi, N.; Gou, J.; Zhu, T.; Jia, L. A novel intelligently integrated MOF-based ratio fluorescence sensor for ultra-sensitive monitoring of TC in water and food samples. *Food Chem.* **2023**, *405*, 134899. [CrossRef]
17. Zhang, Z.; Zhang, Y.; Jayan, H.; Gao, S.; Zhou, R.; Yosri, N.; Zou, X.; Guo, Z. Recent and emerging trends of metal-organic frameworks (MOFs)-based sensors for detecting food contaminants: A critical and comprehensive review. *Food Chem.* **2024**, *448*, 139051. [CrossRef]
18. Marimuthu, M.; Arumugam, S.S.; Sabarinathan, D.; Li, H.; Chen, Q. Metal organic framework based fluorescence sensor for detection of antibiotics. *Trends Food Sci. Technol.* **2021**, *116*, 1002–1028. [CrossRef]
19. Li, S.; Dai, Y.; Ye, P.; Liu, F.; Li, L.; Zhang, D.; Zhu, M. Hierarchical porous MOF/CTF hybrid frameworks used as protection against acidic harmful gases. *Chem. Eng. J.* **2024**, *491*, 152035. [CrossRef]

20. Zhao, W.; Peng, J.; Wang, W.; Liu, S.; Zhao, Q.; Huang, W. Ultrathin two-dimensional metal-organic framework nanosheets for functional electronic devices. *Coord. Chem. Rev.* **2018**, *377*, 44–63. [CrossRef]
21. Muzaffar, A.; Ahamed, M.B.; Deshmukh, K.; Thirumalai, J. A review on recent advances in hybrid supercapacitors: Design, fabrication and applications. *Renew. Sustain. Energy Rev.* **2019**, *101*, 123–145. [CrossRef]
22. Liu, Q.; Guo, Z.; Wang, C.; Guo, S.; Xu, Z.; Hu, C.; Liu, Y.; Wang, Y.; He, J.; Wong, W. A Cobalt-Based Metal-Organic Framework Nanosheet as the Electrode for High-Performance Asymmetric Supercapacitor. *Adv. Sci.* **2023**, *10*, e2207545. [CrossRef] [PubMed]
23. Pan, Y.; Yan, S.; Liu, Y.; Tian, Z.; Li, D.; Chen, Y.; Guo, L.; Wang, Y. Significantly enhanced electrochemical performance of 2D Ni-MOF by carbon quantum dot for high-performance supercapacitors. *Electrochim. Acta* **2022**, *422*, 140560. [CrossRef]
24. Xu, X.; Lu, H.; Xu, D.; Zhou, P.; Ying, Y.; Li, L.; Liu, Y. Oxygen-rich vacancies CuCoLDH with 1D/2D nanoarray structure for high performance asymmetric supercapacitor. *Appl. Surf. Sci.* **2023**, *614*, 156174. [CrossRef]
25. Jia, H.; Lu, S.; Shin, S.H.R.; Sushko, M.L.; Tao, X.; Hummel, M.; Thallapally, P.K.; Liu, J.; Gu, Z. In situ anodic electrodeposition of two-dimensional conductive metal-organic framework@nickel foam for high-performance flexible supercapacitor. *J. Power Sources* **2022**, *526*, 231163. [CrossRef]
26. Bai, W.; Li, S.; Ma, J.; Cao, W.; Zheng, J. Ultrathin 2D metal–organic framework (nanosheets and nanofilms)-based xD–2D hybrid nanostructures as biomimetic enzymes and supercapacitors. *J. Mater. Chem. A* **2019**, *7*, 9086–9098. [CrossRef]
27. Pomerantseva, E.; Gogotsi, Y. Two-dimensional heterostructures for energy storage. *Nat. Energy* **2017**, *2*, 17089. [CrossRef]
28. Lu, X.; Zhang, F.; Sun, Y.; Yu, K.; Guo, W.; Qu, F. A 2D/2D NiCo-MOF/Ti_3C_2 heterostructure for the simultaneous detection of acetaminophen, dopamine and uric acid by differential pulse voltammetry. *Dalton Trans.* **2021**, *50*, 16593–16600. [CrossRef]
29. Ehrnst, Y.; Ahmed, H.; Komljenovic, R.; Massahud, E.; Shepelin, N.A.; Sherrell, P.C.; Ellis, A.V.; Rezk, A.R.; Yeo, L.Y. Acoustotemplating: Rapid Synthesis of Freestanding Quasi-2D MOF/Graphene Oxide Heterostructures for Supercapacitor Applications. *J. Mater. Chem. A* **2022**, *10*, 7058–7072. [CrossRef]
30. Rajak, R.; Saraf, M.; Mobin, S.M. Robust heterostructures of a bimetallic sodium–zinc metal–organic framework and reduced graphene oxide for high-performance supercapacitors. *J. Mater. Chem. A* **2019**, *7*, 1725–1736. [CrossRef]
31. Zhang, X.; Yang, S.; Lu, W.; Lei, D.; Tian, Y.; Guo, M.; Mi, P.; Qu, N.; Zhao, Y. MXenes induced formation of Ni-MOF microbelts for high-performance Supercapacitors. *J. Colloid Interface Sci.* **2021**, *592*, 95–102. [CrossRef]
32. Shao, S.; Zhang, L.; Liu, W.; Zhang, Y.; Shen, X.; Nie, Y.; Yang, W.; Liu, H.; Li, S.; Li, S. Synthesis of Hierarchical Porous MOFs via Ligand Thermolysis for High-Performance Supercapacitor. *J. Inorg. Organomet. Polym. Mater.* **2022**, *32*, 4412–4421. [CrossRef]
33. Li, S.; Yang, K.; Ye, P.; Ma, K.; Zhang, Z.; Huang, Q. Three-dimensional porous carbon/Co_3O_4 composites derived from graphene/Co-MOF for high performance supercapacitor electrodes. *Appl. Surf. Sci.* **2020**, *503*, 144090. [CrossRef]
34. Li, S.; Yang, K.; Ye, P.; Jiang, H.; Zhang, Z.; Huang, Q.; Wang, L. Hierarchical interpenetrating rHGO-decorated $NiCo_2O_4$ nanowires architectures for high-performance supercapacitors. *Appl. Surf. Sci.* **2019**, *473*, 326–333. [CrossRef]
35. Lan, K.; Wei, Q.; Wang, R.; Xia, Y.; Tan, S.; Wang, Y.; Elzatahry, A.; Feng, P.; Mai, L.; Zhao, D. Two-Dimensional Mesoporous Heterostructure Delivering Superior Pseudocapacitive Sodium Storage via Bottom-Up Monomicelle Assembly. *J. Am. Chem. Soc.* **2019**, *141*, 16755–16762. [CrossRef]
36. Babu, R.S.; Vinodh, R.; de Barros, A.; Samyn, L.; Prasanna, K.; Maier, M.; Alves, C.; Kim, H.-J. Asymmetric supercapacitor based on carbon nanofibers as the anode and two-dimensional copper cobalt oxide nanosheets as the cathode. *Chem. Eng. J.* **2019**, *366*, 390–403. [CrossRef]
37. Ma, Y.; Gao, G.; Su, H.; Rong, H.; Lai, L.; Liu, Q. A Cu_4 cluster-based MOF as a supercapacitor electrode material with ultrahigh capacitance. *Ionics* **2021**, *27*, 1699–1707. [CrossRef]
38. Liu, S.; Gordiichuk, P.; Wu, Z.-S.; Liu, Z.; Wei, W.; Wagner, M.; Mohamed-Noriega, N.; Wu, D.; Mai, Y.; Herrmann, A.; et al. Patterning two-dimensional free-standing surfaces with mesoporous conducting polymers. *Nat. Commun.* **2015**, *6*, 8817. [CrossRef] [PubMed]
39. Zou, Y.; Zhou, X.; Ma, J.; Yang, X.; Deng, Y. Recent advances in amphiphilic block copolymer templated mesoporous metal-based materials: Assembly engineering and applications. *Chem. Soc. Rev.* **2020**, *49*, 1173–1208. [CrossRef]
40. Hang, X.; Xue, Y.; Cheng, Y.; Du, M.; Du, L.; Pang, H. From Co-MOF to CoNi-MOF to Ni-MOF: A Facile Synthesis of 1D Micro-/Nanomaterials. *Inorg. Chem.* **2021**, *60*, 13168–13176. [CrossRef]
41. Liu, F.; Ye, P.; Cheng, Q.; Zhang, D.; Nie, Y.; Shen, X.; Zhu, M.; Xu, H.; Li, S. By introducing multiple hydrogen bonds endows MOF electrodes with an enhanced structural stability. *Inorg. Chem.* **2024**, *63*, 14630–14640. [CrossRef]
42. Nicks, J.; Zhang, J.; Foster, J.A. Tandem catalysis by ultrathin metal–organic nanosheets formed through post-synthetic functionalisation of a layered framework. *Chem. Commun.* **2019**, *55*, 8788–8791. [CrossRef] [PubMed]
43. Rad, M.; Dehghanpour, S. ZnO as an efficient nucleating agent and morphology template for rapid, facile and scalable synthesis of MOF-46 and ZnO@MOF-46 with selective sensing properties and enhanced photocatalytic ability. *RSC Adv.* **2016**, *6*, 61784–61793. [CrossRef]
44. Zhao, L.; Chen, J.; Xiong, N.; Bai, Y.; Yilihamu, A.; Ma, Q.; Yang, S.; Wu, D.; Yang, S.-T. Carboxylation as an effective approach to improve the adsorption performance of graphene materials for Cu^{2+} removal. *Sci. Total. Environ.* **2019**, *682*, 591–600. [CrossRef]

45. Cui, M.; Park, S.-J.; Kim, S. Carboxylated group effect of graphene oxide on capacitance performance of Zr-based metal organi framework electrodes. *J. Inorg. Organomet. Polym. Mater.* **2021**, *31*, 1939–1945. [CrossRef]
46. Yuan, Y.; Gao, X.; Wei, Y.; Wang, X.; Wang, J.; Zhang, Y.; Gao, C. Enhanced desalination performance of carboxyl functionalize graphene oxide nanofiltration membranes. *Desalination* **2017**, *405*, 29–39. [CrossRef]
47. Guo, S.; Raya, J.; Ji, D.; Nishina, Y.; Ménard-Moyon, C.; Bianco, A. Is carboxylation an efficient method for graphene oxid functionalization? *Nanoscale Adv.* **2020**, *2*, 4085–4092. [CrossRef] [PubMed]
48. Li, X.; Ding, R.; Shi, W.; Xu, Q.; Liu, E. Ternary Ni-Co-F nanocrystal-based supercapacitors. *Chem.-Eur. J.* **2017**, *23*, 6896–690 [CrossRef]
49. Kong, M.; Wang, Z.; Wang, W.; Ma, M.; Liu, D.; Hao, S.; Kong, R.; Du, G.; Asiri, A.M.; Yao, Y.; et al. NiCoP nanoarray: A superio pseudocapacitor electrode with high areal capacitance. *Chem.-Eur. J.* **2017**, *23*, 4435–4441. [CrossRef]
50. Cheng, H.; Xiao, R.; Bian, H.; Li, Z.; Zhan, Y.; Tsang, C.K.; Chung, C.; Lu, Z.; Li, Y.Y. Periodic porous silicon thin films wit interconnected channels as durable anode materials for lithium ion batteries. *Mater. Chem. Phys.* **2014**, *144*, 25–30. [CrossRef]
51. Wang, J.; Zhong, Q.; Xiong, Y.; Cheng, D.; Zeng, Y.; Bu, Y. Fabrication of 3D Co-doped Ni-based MOF hierarchical micro-flower as a high-performance electrode material for supercapacitors. *Appl. Surf. Sci.* **2019**, *483*, 1158–1165. [CrossRef]
52. Gittins, J.W.; Balhatchet, C.J.; Chen, Y.; Liu, C.; Madden, D.G.; Britto, S.; Golomb, M.J.; Walsh, A.; Fairen-Jimenez, D.; Dutton, S.E et al. Insights into the electric double-layer capacitance of two-dimensional electrically conductive metal–organic frameworks. *Mater. Chem. A* **2021**, *9*, 16006–16015. [CrossRef]
53. Ramachandran, R.; Zhao, C.; Luo, D.; Wang, K.; Wang, F. Synthesis of copper benzene-1, 3, 5-tricarboxylate metal organi frameworks with mixed phases as the electrode material for supercapacitor applications. *Appl. Surf. Sci.* **2018**, *460*, 33–3 [CrossRef]
54. Rajak, R.; Saraf, M.; Kumar, P.; Natarajan, K.; Mobin, S.M. Construction of a Cu-based metal–organic framework by employing a mixed-ligand strategy and its facile conversion into nanofibrous CuO for electrochemical energy storage applications. *Inorg Chem.* **2021**, *60*, 16986–16995. [CrossRef]
55. Zou, J.; Xie, D.; Xu, J.; Song, X.; Zeng, X.; Wang, H.; Zhao, F. Rational design of honeycomb Ni-Co LDH/graphene composite for remarkable supercapacitor via ultrafast microwave synthesis. *Appl. Surf. Sci.* **2022**, *571*, 151322. [CrossRef]
56. Wang, J.; Polleux, J.; Lim, J.; Dunn, B. Pseudocapacitive contributions to electrochemical energy storage in TiO_2 (anatase nanoparticles. *J. Phys. Chem. C* **2007**, *111*, 14925–14931. [CrossRef]
57. Deng, Y.; Wang, X.; Wang, Z.; Wang, X.; Li, Z.; Wang, L.; Zhou, C.; Chen, D.; Luo, Y. Yolk–shell structured nickel cobalt sulfide and carbon nanotube composite for high-performance hybrid supercapacitors. *Energy Fuels* **2021**, *35*, 5342–5351. [CrossRef]
58. Mao, F.; Li, Y.; Zou, Z.; Huang, B.; Yang, J.; Yao, J. Zn^{2+} storage performance and structural change of orthorhombic V_2O_5 nanowires as the cathode material for rechargeable aqueous zinc-ion batteries. *Electrochim. Acta* **2021**, *397*, 139255. [CrossRef]
59. Ninawe, P.; Gupta, K.; Ballav, N. Chemically integrating a 2D metal–organic framework with 2D functionalized graphene. *Inorg. Chem.* **2021**, *60*, 19079–19085. [CrossRef] [PubMed]
60. Li, W.; Ding, K.; Tian, H.; Yao, M.; Nath, B.; Deng, W.; Wang, Y.; Xu, G. Conductive metal–organic framework nanowire array electrodes for high-performance solid-state supercapacitors. *Adv. Funct. Mater.* **2017**, *27*, 1702067. [CrossRef]
61. Liu, J.; Zhou, Y.; Xie, Z.; Li, Y.; Liu, Y.; Sun, J.; Ma, Y.; Terasaki, O.; Chen, L. Inside cover: Conjugated copper–catecholate framework electrodes for efficient energy storage. *Angew. Chem. Int. Ed.* **2020**, *59*, 962. [CrossRef]
62. Wang, J.; Zhong, Q.; Zeng, Y.; Cheng, D.; Xiong, Y.; Bu, Y. Rational construction of triangle-like nickel-cobalt bimetallic metal-organic framework nanosheets arrays as battery-type electrodes for hybrid supercapacitors. *J. Colloid Interface Sci.* **2019**, *555*, 42–52. [CrossRef] [PubMed]

nanomaterials

MDPI

Article

Wafer-Level Fabrication of Radiofrequency Devices Featuring 2D Materials Integration

Vitor Silva [1,2,3,†], Ivo Colmiais [1,2,3,†], Hugo Dinis [1,2], Jérôme Borme [3], Pedro Alpuim [3,4] and Paulo M. Mendes [1,2,*]

1 CMEMS—Center for Microelectromechanical Systems, University of Minho, Campus de Azurém, 4800-058 Guimarães, Portugal; hdinis@dei.uminho.pt (H.D.)
2 LABBELS—Associate Laboratory, 4710-057 Braga, Portugal
3 INL—International Iberian Nanotechnology Laboratory, Av. Mestre José Veiga, 4715-330 Braga, Portugal; jerome.borme@inl.int (J.B.); pedro.alpuim.us@inl.int (P.A.)
4 Center of Physics, University of Minho, Campus de Gualtar, 4710-057 Braga, Portugal
* Correspondence: paulo.mendes@dei.uminho.pt
† These authors contributed equally to this work.

Abstract

Two-dimensional (2D) materials have been proposed for use in a multitude of applications, with graphene being one of the most well-known 2D materials. Despite their potential to contribute to innovative solutions, the fabrication of such devices still faces significant challenges. One of the key challenges is the fabrication at a wafer-level scale, a fundamental step for allowing reliable and reproducible fabrication of a large volume of devices with predictable properties. Overcoming this barrier will allow further integration with sensors and actuators, as well as enabling the fabrication of complex circuits based on 2D materials. This work presents the fabrication steps for a process that allows the on-wafer fabrication of active and passive radiofrequency (RF) devices enabled by graphene. Two fabrication processes are presented. In the first one, graphene is transferred to a back gate surface using critical point drying to prevent cracks in the graphene. In the second process, graphene is transferred to a flat surface planarized by ion milling, with the gate being buried beneath the graphene. The fabrication employs a damascene-like process, ensuring a flat surface that preserves the graphene lattice. RF transistors, passive RF components, and antennas designed for backscatter applications are fabricated and measured, illustrating the versatility and potential of the proposed method for 2D material-based RF devices. The integration of graphene on devices is also demonstrated in an antenna. This aimed to demonstrate that graphene can also be used as a passive device. Through this device, it is possible to measure different backscatter responses according to the applied graphene gating voltage, demonstrating the possibility of wireless sensor development. With the proposed fabrication processes, a flat graphene with good quality is achieved, leading to the fabrication of RF active devices (graphene transistors) with intrinsic f_T and f_{max} of 14 GHz and 80 GHz, respectively. Excellent yield and reproducibility are achieved through these methods. Furthermore, since the graphene membranes are grown by Chemical Vapor Deposition (CVD), it is expected that this process can also be applied to other 2D materials.

Keywords: graphene; radiofrequency; GFET; antennas; nanofabrication

1. Introduction

Graphene has been widely presented as a wonder material with extraordinary properties, with a wide range of applications and the potential to revolutionize various industries.

One possible application of graphene is electronic circuits, more specifically for RF applications. Graphene's properties originate from its unique honeycomb lattice that is formed by a single sheet of sp^2-hybridized carbon atoms, resulting in a zero-bandgap semiconductor. This allows graphene to have a relatively high conductivity, low resistance, sustain a high current density [1], and possess high carrier mobility that can reach up to 200,000 $cm^2\ V^{-1}s^{-1}$ [2]. Graphene RF transistors have the potential to achieve cutoff frequencies in the hundreds of GHz [3]. It becomes apparent that, by integrating graphene and exploring its properties, it should be possible to improve electrical circuits [1,4,5].

However, there are various limitations in the fabrication of graphene-based devices. An example of this is transistors, whose performance is typically severely degraded. One of the reasons for this is the fact that high-κ gate dielectrics are required to maximize the RF performance. However, it is difficult to grow a high-κ oxide on top of graphene due to the lack of reactive sites for Atomic Layer Deposition (ALD), especially when thin oxide layers are required. Another reason is the high contact resistance between graphene and its metal contact [6].

To improve RF performance, several techniques can be implemented: the use of physical gates to increase the f_{max} of the transistors [7], or complex gate structures designed to reduce parasitic and access resistances [8]. Self-aligned structures are also being developed to minimize access resistance, thereby improving the RF performance of GFETs [9], which is a critical parameter for achieving high-performance graphene RF oscillators.

A comprehensive review on graphene applications in the design of RF building blocks, their performance, and challenges is presented by the authors in [6]. Additionally, this review approaches RF transistor state of the art, detailing applications such as oscillators, multipliers, and mixers. Finally, the review also discusses fabrication techniques and issues, providing a comprehensive overview of graphene for RF applications.

In this work, the design and fabrication of graphene devices is discussed. Two fabrication approaches are presented: a buried bottom-gate graphene device fabrication process without surface planarization, using a critical point dryer to achieve crack-free graphene transfer, and a surface planarization technique employing ion milling prior to graphene transfer. These techniques have the objective of preserving the graphene lattice so that high carrier mobilities can be achieved. In both cases, Chemical Vapor Deposition (CVD)-grown graphene is used, as well as the PMMA-assisted transfer method. The performance of devices produced with both fabrication approaches is assessed and compared, allowing for a better understanding of the performance differences between the two approaches.

To demonstrate the development of graphene RF devices, the surface planarization technique with ion milling prior to graphene transfer is used to fabricate an inverter and a nonlinear block. Finally, to demonstrate the application of the fabrication process on the development of RF electronics, graphene is included as a passive element on an antenna, allowing for a backscatter readout that can be implemented in electronic sensors.

The manuscript is organized as follows. Section 1 is an introduction. Section 2 describes two graphene RF device fabrication techniques. Section 3 contains the experimental validation of the RF graphene transistors fabricated with both the methods proposed in Section 2. Section 4 describes the development of RF circuits, such as an oscillator, with graphene transistors, as well as the development and testing of passive RF graphene components. Section 5 presents this work's conclusions.

2. Methodology

As previously discussed, two approaches to fabricating active devices (RF graphene transistors) and passive devices such as coils, antennas, and capacitors will be presented in this work. First, we present a bottom-gate graphene device fabrication process without surface planarization, along with a detailed discussion of the associated challenges. To achieve a crack-free graphene transfer, a critical point-drying process was developed and employed to dry the samples containing the devices, preserving the graphene lattice characteristics. In the other approach, a surface planarization technique using ion milling prior to graphene transfer was applied, resulting in the successful fabrication of a buried gate RF graphene FET, demonstrating the effectiveness of the planarization process.

The diagram of the two fabrication processes reported in this manuscript can be seen in Figure 1.

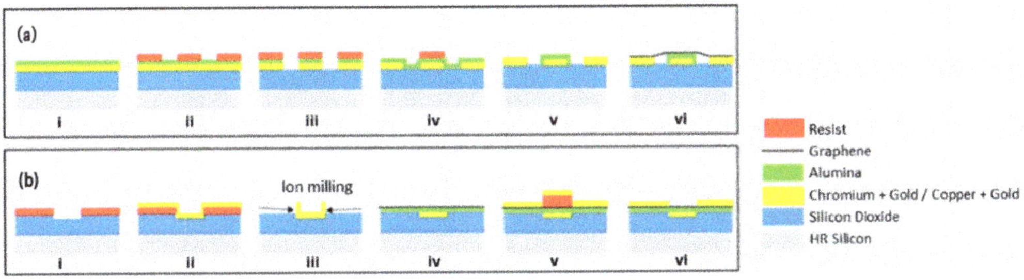

Figure 1. Diagram of the two main processes reported in this work, with and without planarization, (**a**,**b**), respectively. (**a**) (i) high resistivity (HR) silicon wafer with chromium + gold and alumina on top; (ii) e-beam lithography to define source, drain, and gate contacts by ion milling; (iii) device after ion milling; (iv) e-beam lithography to define the gate oxide after the growth of fresh alumina by ALD (after the resist removal by O_2 plasma, and alumina by wet etch); (v) result after the patterning of the gate oxide by ion milling; (vi) final device with graphene. (**b**) (i) device after e-beam lithography and inductively coupled plasma (ICP) reactive ion etching (RIE); (ii) after the chromium gold deposition; (iii) after the lift-off, showing the ears to be removed by ion milling, to planarize the device; (iv) after the graphene transfer; (v) after the e-beam lithography and copper + gold deposition to define the source/drain electrodes by lift-off; (vi) final device.

2.1. Graphene Device Fabrication Process with CPD

The fabrication of the bottom-gate devices without planarization began with an HR silicon wafer, used to minimize parasitic capacitances from the substrate, on which 500 nm of SiO_2 was grown via plasma-enhanced chemical vapor deposition (PECVD). Subsequently, 3 nm of chromium and 97 nm of gold were deposited by sputtering. To define the contacts using ion milling, 10 nm of alumina was sputtered on top of the gold to facilitate the removal of the photoresist. After alumina deposition, the sample was prepared for electron beam (e-beam) lithography. The wafer was coated with AR-N 7520.18, and e-beam lithography was performed to pattern the source, drain, and gate contacts. The contacts were then patterned using ion milling at an angle of 130 degrees, followed by 165 degrees (relative to the surface) to remove redeposited metal "ears". The photoresist was stripped using an oxygen plasma, and an aluminum etchant (Fujifilm 16:1:1:2 aluminum etch) was employed to remove the alumina, resulting in a clean surface devoid of resist residues. These steps are shown in Figure 2.

Next, 10 nm of Al_2O_3 was deposited by ALD to serve as the gate dielectric. The dielectric was patterned using ion milling after an e-beam lithography with AR-N 7520.18. The photoresist was then removed using an oxygen plasma, as illustrated in Figures 3 and 4.

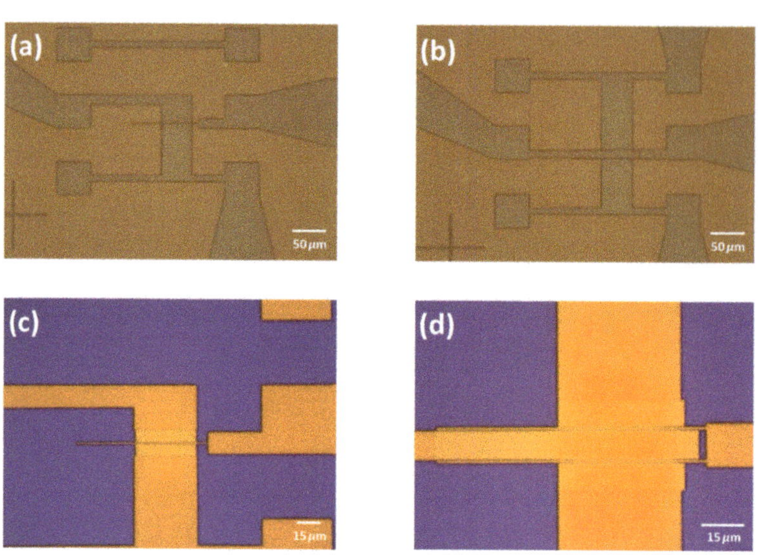

Figure 2. First steps of the fabrication process with (**a**,**b**) showing an optical image of the lithography and (**c**,**d**) showing the optical micrograph after the ion milling and the resist removal by an O_2 plasma (intermediate step between (iii–iv) in Figure 1).

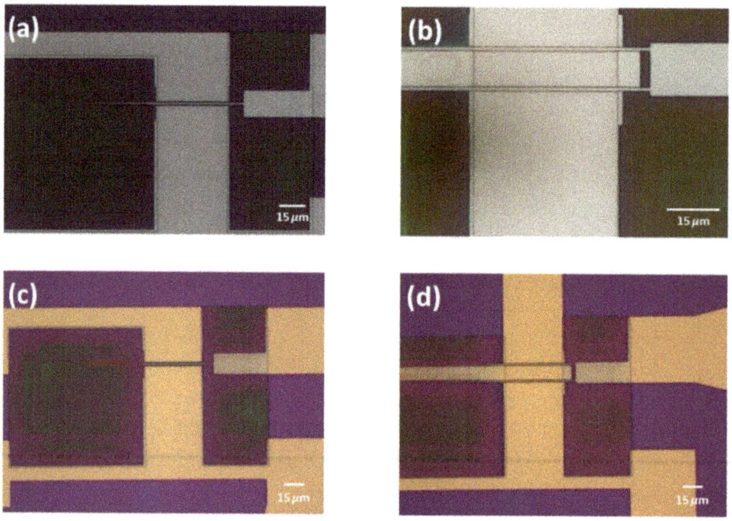

Figure 3. (**a**,**b**) An optical image of the lithography performed to pattern the gate dielectric (alumina), and (**c**,**d**) show the optical image after the ion milling. To note, since the gate is very thin, some e-beam resist was left to act as anchor of the e-beam resist responsible for the patterning of the gate dielectric.

CVD graphene grown on copper foil was transferred onto the top of the structures using a PMMA-assisted wet transfer, as reported by our group in [10]. Shortly prior to the transfer, an O_2 plasma treatment was applied to the back side of the copper foil (which contains the graphene protected with the PMMA) to remove the graphene from that side. The copper foil was then etched in an iron chloride solution (0.5 M). Subsequently, the graphene-PMMA stack was transferred to an HCl (2%) solution to remove iron chloride contaminants and then to water before the final transfer to the substrate. Before this, the substrate surface was dehydrated and primed using a vapor prime oven to promote the adhesion of the graphene to the final substrate. After the transfer, the graphene was dried at room temperature. The PMMA was then removed using acetone. Finally, the graphene was patterned using an O_2 plasma beam following optical lithography with AZ1505. The resist was removed using acetone overnight. In Figure 5, it is possible to observe the graphene lying in the graphene channel.

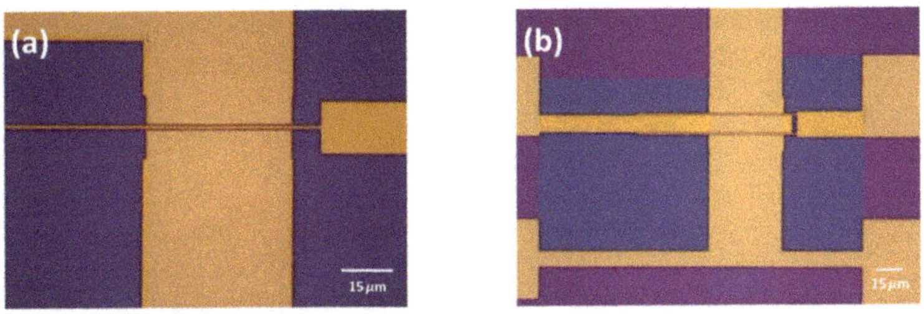

Figure 4. (**a**,**b**) Optical photograph of the device after the resist removal with the oxygen plasma.

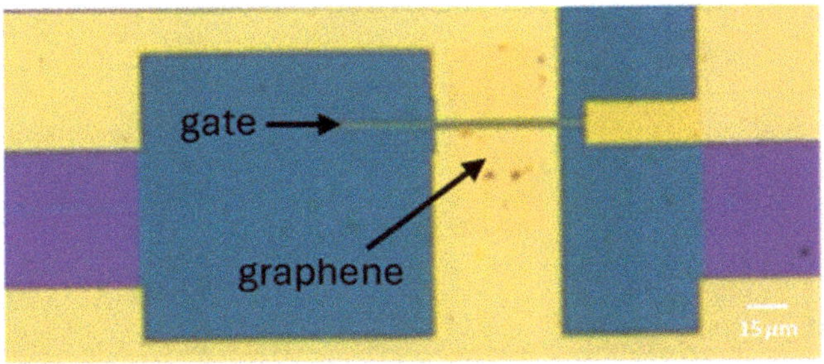

Figure 5. Optical image of the graphene lying in the channel of the device.

Due to the small gaps between the source and drain contacts (as shown in Figure 6, mandatory for RF devices), the drying process is critically important. Two approaches were tested with the continuous graphene film after patterning (using O_2 plasma, as it was discovered that graphene only breaks after patterning, i.e., after the removal of the photoresist, not after the removal of the PMMA used for transfer).

In the first approach, the sample was dried in air. In the second approach, critical point drying was used. In the first approach, the surface tension of acetone (during the drying process) resulted in damage to the graphene at some sites, making it unsuitable for RF device fabrication (see Figure 6c,d). For thicknesses above approximately 500 nm, the graphene tends to conform to the surface of the contacts rather than remaining suspended (see Figure 6b), which mitigates the impact of the drying process. This observation supports the use of coplanar structures in sensor fabrication, where graphene on the sidewalls enhances sensor sensitivity [11]. The performance of the fabricated devices is shown in the next section.

In contrast, the second approach, using critical point drying, allowed the graphene to remain suspended due to the uniform drying process (see Figure 6e) and Figure 6f.

Since the use of critical point drying (CPD) poses significant challenges in handling delicate samples, increases the risk of contamination, and as will be shown later, results in devices with limited performance, a new method is proposed. Furthermore, because the graphene must remain immersed in ethanol after the graphene transfer and cannot be allowed to dry at any stage, the transfer and immersion steps become complex and increase the risk of damaging the graphene. To address these issues, a planar bottom-gate topology was developed, offering greater robustness, compatibility with standard lithographic techniques, and improved scalability.

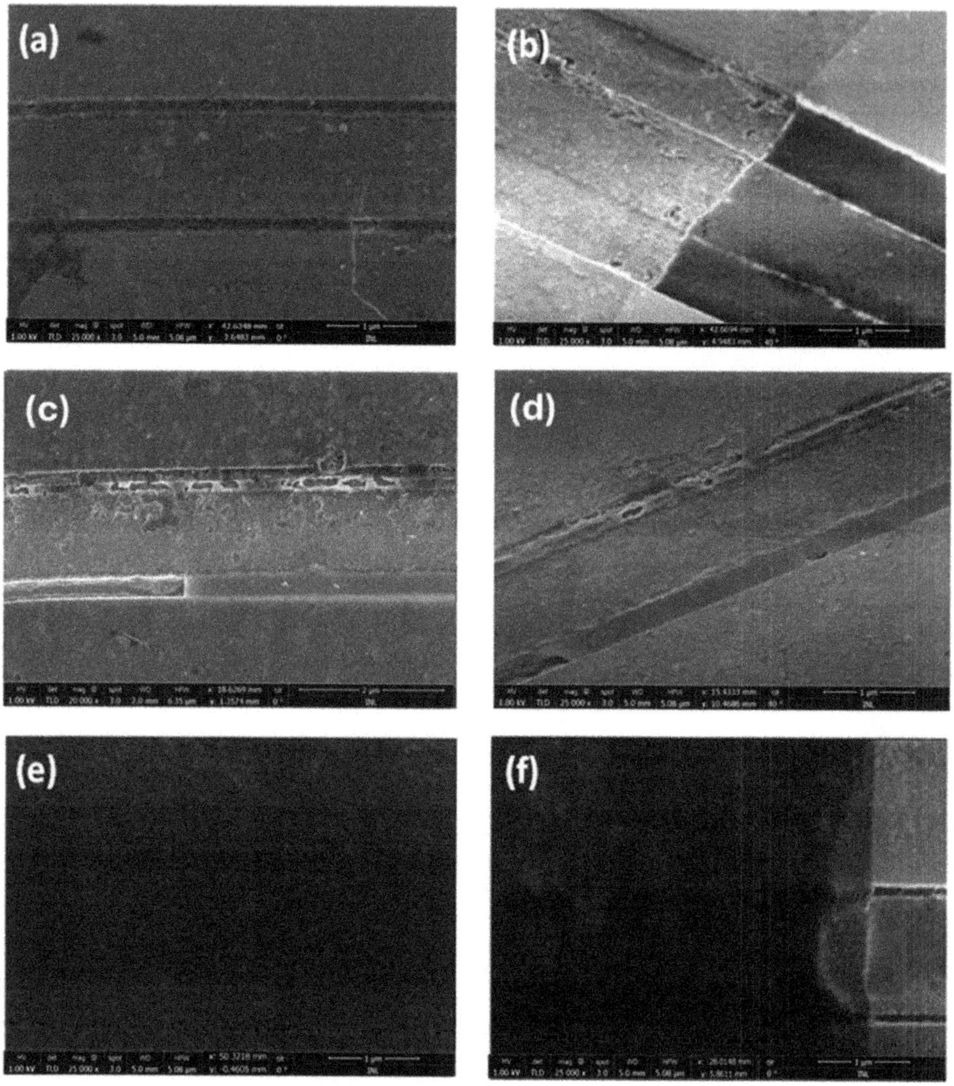

Figure 6. SEM images of graphene after the transfer at 1 kV. (**a**) After the transfer without the patterning, (**b**) after the patterning with graphene following the topography of the contacts, (**c**,**d**) suspended graphene patterned and dried in air, evidencing the cracks, and (**e**,**f**) after the patterning and dried with critical point dryer.

2.2. Graphene Device Fabrication Process with Surface Planarization

The gate contact was buried within the SiO_2, and its surface was partially planarized using ion milling at a grazing angle, leading to a reduction in topography by a scale factor of 10. In this way, the graphene is transferred onto a flat surface, avoiding mechanical stress or rupture and thus preserving its characteristic and essential properties—such as high electron mobility—that are crucial for the optimal performance of RF devices. Initially, 500 nm of SiO_2 was grown on a high-resistivity silicon wafer via PECVD. The gate contact was defined using e-beam lithography, involving the sequential spin coating of AR-P 617.03 and AR-P 6200.09 resists. These resists were selected to simplify the lift-off process for the gate metal. An Inductive Coupled Plasma Reactive Ion Etching (ICP RIE) was performed to open the SiO_2 trenches without compromising the resists for the lift-off process. Without removing the resist mixture, 3 nm of chromium and 67 nm of gold were sputtered onto the sample. The metal was subsequently lifted off using a solution of *N*-ethylpyrrolidone (NEP) with ultrasonic agitation. Due to the lift-off process, lift-off ears were observed, as shown in Figure 7a. These lift-off ears were removed by ion milling at a 174-degree beam angle (relative to the surface, resulting in a flat surface suitable for graphene transfer (see Figure 7b,c). Note that in (c), it is possible to see only small ears around 4 nm, which could be assumed as a flat surface for graphene.

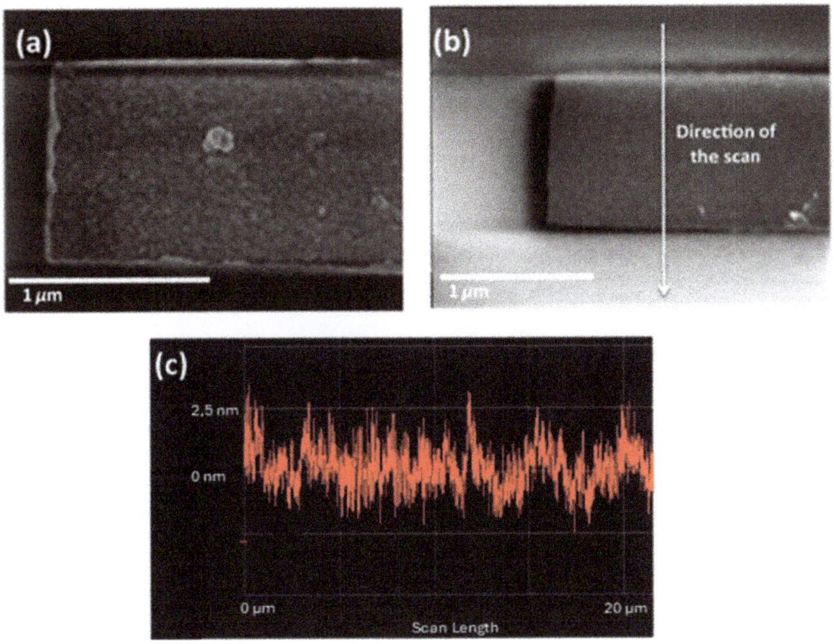

Figure 7. SEM images of the gate structures showing the metal ears (**a**) and after the planarization (**b**). (**c**) Profilometer measurement of the gate after the planarization.

After planarization, 10 nm of alumina were deposited using ALD. Standard optical lithography was then employed to selectively remove the alumina via wet etching (Fujifilm 16:1:1:2 aluminum etch) to expose areas for probe contact. Graphene was subsequently transferred onto the surface and patterned with the previously described method. The source/drain electrodes were defined using e-beam lithography with a Poly (methyl methacrylate) (PMMA) bilayer (AR-P 617.03 + AR-P 679.04). Following this, 5 nm of copper and 50 nm of gold were sputtered and then lifted off using acetone. Copper was chosen as the adhesion layer because sputtered chromium, when used with gold, exhibited higher contact resistance compared to copper. The final device can be observed in Figure 8. It is important to note that after the graphene transfer in both fabrication processes, the quality of the graphene was assessed via Raman spectroscopy, as shown in Figure 9.

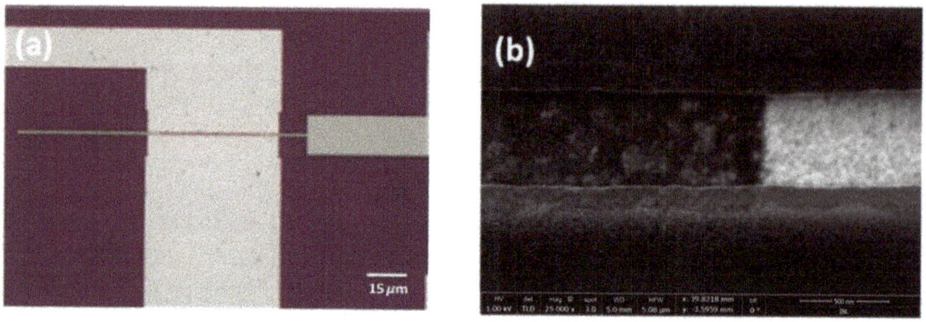

Figure 8. (**a**) Optical image of the final device, and (**b**) SEM image of the final device showing graphene without cracks.

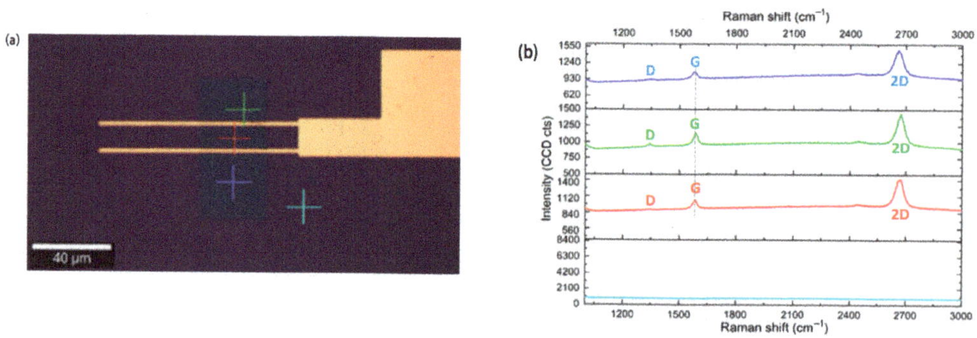

Figure 9. Raman spectra of the monocrystalline graphene used in the fabricated devices, (**a**) Raman spectra sites, and (**b**) Raman spectra acquired with a 532 nm laser (0.5 mW) using a confocal Raman microscope and a 50× objective. D: defect. G: graphene. 2D: 2D peak. It is possible to evaluate the correct patterning since, out of the graphene channel, there is no graphene signature.

3. Performance Assessment of the Proposed Wafer-Level Fabrication

The proposed fabrication methods were employed to produce sets of GFETs. The performance assessment of these transistors is reported in this section of the manuscript. Graphene purity assessment is performed by resorting to Raman spectroscopy. Then, the electrical performance of the transistors is verified for low- and high-frequency signals.

3.1. Graphene Quality Assessment

In the previously reported fabrication processes, monocrystalline graphene was used to achieve a high yield in the fabricated devices. The graphene was carefully transferred onto the substrates to maintain its crystalline integrity. To ensure the quality and continuity of the graphene layer, Raman spectroscopy was employed. The Raman spectra of the graphene layer can be observed in Figure 9.

As observed in the Raman spectra, a blue shift occurs, specifically with the G peak shifting to a higher wavenumber. This shift indicates the presence of p-type doping [12]. The p-doping may originate from the fabrication process (PMMA contaminants or photoresist [10]) or from contaminants in the ambient environment that interact with the graphene surface [13]. Furthermore, the Raman spectra confirm that the graphene is predominantly monolayer (evidenced by the $I_{2D}/I_G > 2$), although defects are present, as evidenced by the appearance of the D peak. The successful patterning of the graphene is demonstrated by the absence of graphene outside the designated channel region, as seen both in the corresponding images and Raman spectra. The graphene mobility was extracted through the transfer curve fitting method, being 900 cm^2 V^{-1} s^{-1} for holes and 1100 cm^2 V^{-1} s^{-1} for electrons [14].

3.2. *Low-Frequency Electrical Performance*

3.2.1. Graphene Device Fabrication Process Without Surface Planarization (CPD)

The DC measurements were performed to obtain the I_{DS} and the respective g_m. To do this, two DC sources were required, one connected to the source and drain contacts and another to the gate and source contacts, as shown in Figure 10. The measurements were performed by measuring the I_{DS}, changing one of the DC sources, and keeping the other constant. The results for a device with W = 34 μm and L = 1.100 μm are shown in Figure 11. The maximum extracted g_m is 4 μS.

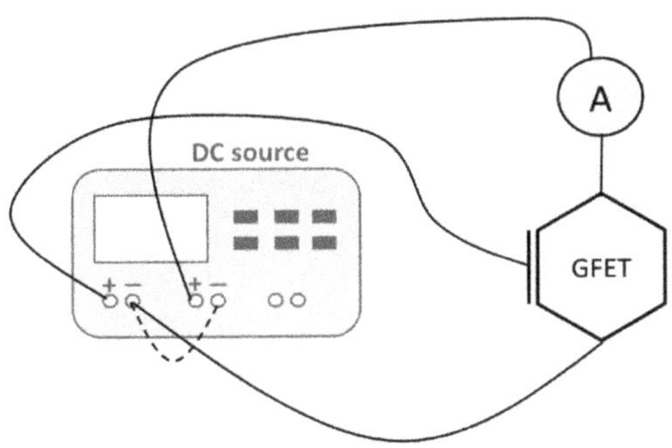

Figure 10. Measurement setup to extract the transfer curve of GFET transistors.

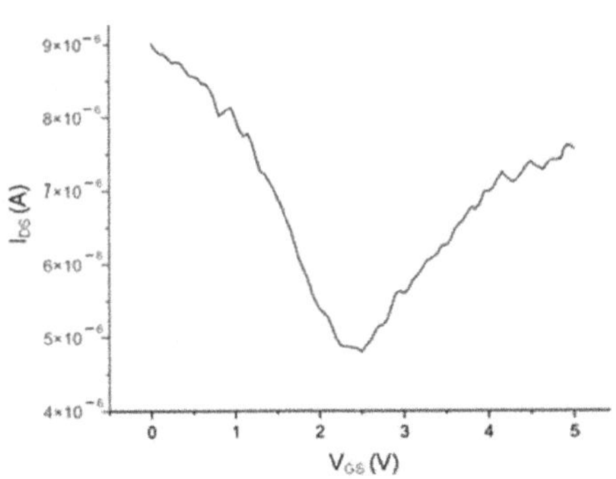

Figure 11. I_{DS} vs. V_{GS} and g_m characteristic curves of a device with W = 34 μm and L = 1.100 μm (0.95 μm of gate channel and 0.075 μm of drain/gate and source/gate overlap) measured at a V_{DS} of 100 mV.

3.2.2. Planar Buried Bottom-Gate Topology Improved

The low-frequency electrical properties of GFETs fabricated with the planar buried bottom-gate topology were evaluated with the same procedure as the previously reported CPD-fabricated GFTETs. The data derived from these measurements (transfer curves) are illustrated in the following figure for a device with L = 1.101 μm and W = 33.82 μm, similar dimensions to the previously shown device in Figure 12.

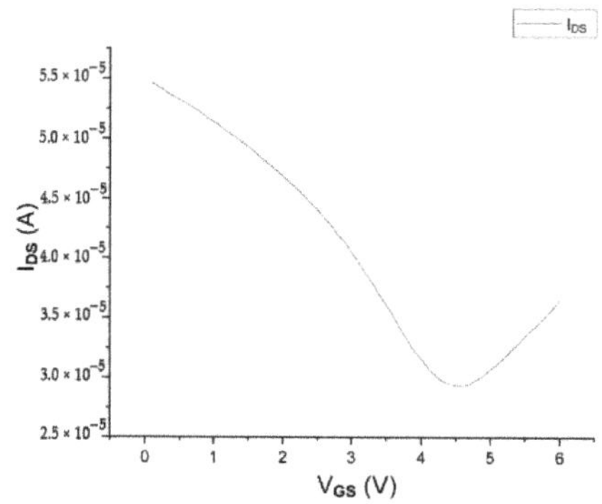

Figure 12. Transfer curve of a device with L = 1.102 μm and W = 33.82 μm. V_{GS} vs. I_{DS} curve for a V_{DS} of 10 mV.

The Dirac voltage of the device is observed at approximately 4.5 V (positive voltage), indicating p-type doping, as previously suggested during the Raman evaluation. A g_m of 10 μS was obtained, which is bigger than the one achieved in 3.2.1, showing an increase in the device performance for DC applications. To note that due to the different graphene sample used to fabricate the devices, the p-doping is more prominent in the planar buried bottom-gate topology process, as shown in the shift of the Dirac point in Figure 12 when compared with Figure 11.

3.3. High-Frequency Electrical Performance

The electrical performance is evaluated through the extraction of the figure of merit (FOM) for graphene field-effect transistors (GFETs). This setup incorporates two bias tees to separate direct current (DC) and radio frequency (RF) signals, enabling the simultaneous application of a DC bias voltage and an RF signal to the device. A DC power supply is employed to provide the necessary gate and drain bias voltages, thereby biasing the GFET into the desired operational region. A E5071C VNA (Keysight, Santa Rosa, CA, USA) is used to generate and measure high-frequency RF signals, allowing for the determination of the device's scattering parameters (S-parameters), which are critical for the calculation of the transistor's FOM. Additionally, a control computer interface is employed to regulate the gate's DC voltage and to facilitate the extraction and analysis of the S-parameters. The measurement setup is represented in Figure 13.

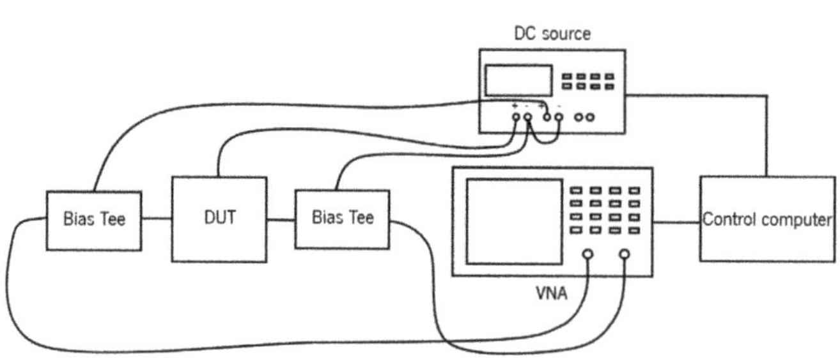

Figure 13. Measurement setup used to extract the FOMs of the GFETs.

3.3.1. Graphene Device Fabrication Process with CPD

To evaluate the performance of the devices fabricated using the CPD method, extrinsic RF characterization was performed. Figure 14 illustrates measured S-parameters (a) and the extracted FOMs (b). In this device, an extrinsic cut-off frequency (fT) of 5.5 GHz and a maximum oscillation frequency (fmax) of 0.7 GHz, measured at a VDS of 3 V and a VGS of 5 V, were demonstrated. These values were obtained without employing a global back gate, suggesting a high level of p-doping in the graphene membrane. This fabrication method achieved a high yield of operational devices, with approximately 70% exhibiting RF functionality.

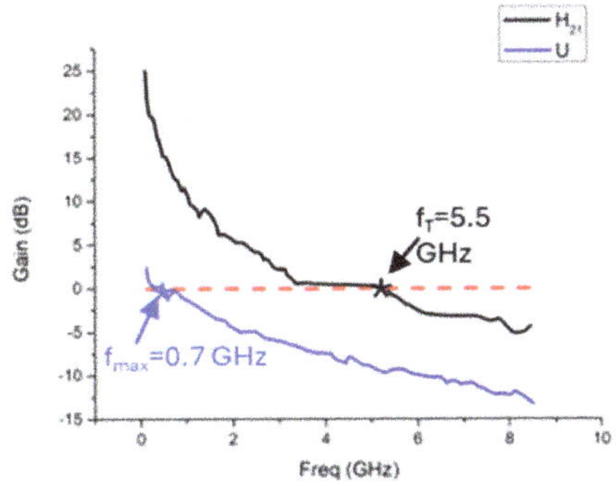

Figure 14. Gain (dB) vs. Freq (GHz) of a device with W = 34 μm and L = 1.100 μm and a La (access length) of 85 nm and the extracted fT and fmax.

3.3.2. Planar Buried Bottom-Gate Topology

Considering the DC performance of the devices, the RF performance was also evaluated. Notably, de-embedded structures were fabricated to extract both the intrinsic fT and extrinsic fmax, isolating the contributions from parasitic elements external to the device that significantly impact its performance. The open structure consisted of a device similar to the fabricated transistors but without graphene in the channel. The short structure, in contrast, was fabricated with all electrodes connected. Figure 15 shows the RF performance of two distinct devices after the de-embedding, showing the extrinsic and intrinsic performance of the transistors. As concluded earlier, and as expected, the device with a small La exhibited a better performance than the one without La.

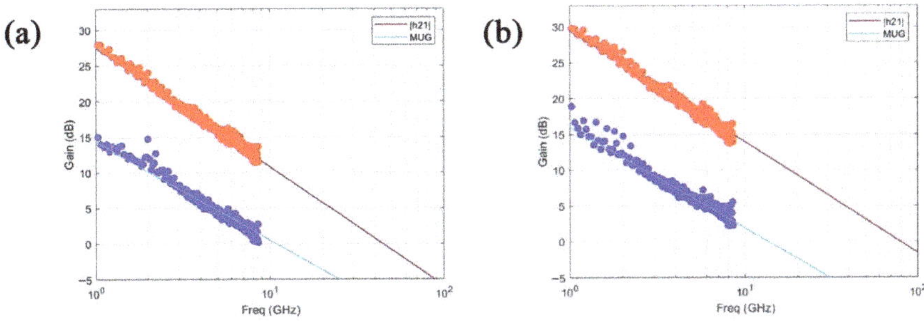

Figure 15. RF performance of the de-embedded devices. H21 and MUG of a device with an L = 1.101 μm (**a**,**b**) of a device with a L = 1.136 μm and a La of 54.35 nm. The f_{max} and f_T of (**a**) are 11 GHz and 44 GHz, respectively, and in (**b**) are 14 GHz and 80 GHz, respectively.

4. Performance Assessment of Fabricated RF Building Blocks

To assess the viability of the proposed methodology, both active and passive devices were fabricated and integrated into more complex circuits, demonstrating the potential of this process for use in systems capable of performing advanced functions.

4.1. Performance Assessment of Fabricated Active Devices

The fabricated devices underwent validation through integration into various circuits built with active devices to demonstrate their suitability for real-world applications. The circuits were selected to cover a representative range of functionalities, from basic logic operations to more complex nonlinear applications. For instance, standard inverters were implemented to evaluate the devices' switching capabilities and overall performance in digital applications. Additionally, the devices were tested in nonlinear circuits, such as mixers and frequency doublers, which are critical in RF and communication systems. Mixers were employed to analyze the devices' behavior in frequency conversion, and frequency doublers were tested to provide insights into harmonic generation and efficiency under nonlinear conditions. These tests helped characterize the devices under different operating conditions, providing a comprehensive understanding of their potential for practical use.

4.1.1. Transistor Integrated to Perform Digital Operations

The graphene inverters, which serve as the fundamental building blocks for RF graphene oscillators, were fabricated and validated to evaluate their RF performance and suitability for integration. To test the inverter, a square wave with a 1 V amplitude swing centered at 5.7 V (chosen because of the Dirac voltages since V_{Dirac} are far from 0 V) was applied to the inverter's input (V_{in}). The schematic of the inverter is shown in Figure 16a, while the fabricated device is depicted in Figure 16b. The inverter was powered by a supply voltage (V_{DD}) of 8 V. For the applied V_{in}, the measured output voltage (V_{out}) was 500 mV, corresponding to an inverter gain (A_v) of 0.5. The output signal is centered at 0 V due to the use of a bias tee, which isolates and recovers the AC component of the output. Figure 16c presents the oscilloscope capture, displaying both the input and output signals, further confirming the inverter's operation and performance.

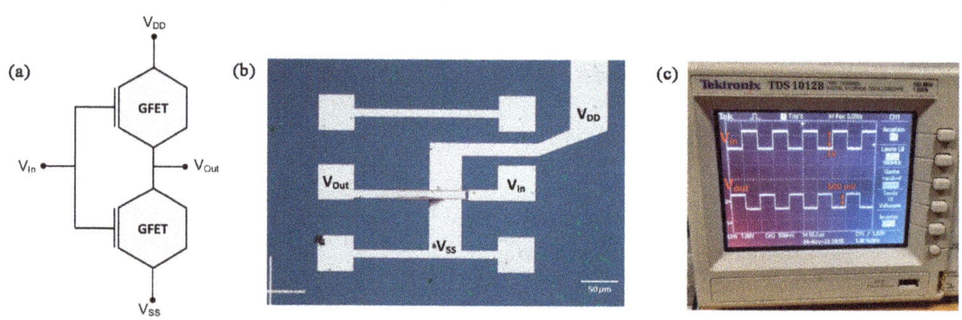

Figure 16. (**a**) Inverter schematic with GFET. (**b**) Equivalence of the fabricated device with the schematic shown in (**a**). (**c**) Measurement of a graphene inverter with a V_{DD} of 8V showing a V_{out} of 500 mV of amplitude swing (channel 2) for a V_{in} of 1 V swing centered in 5.7 V.

Despite the favorable f_T and f_{max} reported earlier, the measured inverter did not exhibit a gain greater than 1 (A_v = 0.5), a criterion essential for oscillation according to Barkhausen's stability criterion. This limitation may arise from the selected device or the design of the inverter itself. Efforts were made to measure the fabricated ring oscillator to validate its functionality; however, no output was observed when using the RF probes. To ensure proper operation of the ring oscillators, it is necessary to increase the gain of the inverter. However, due to the high p-doping observed in the fabricated transistors, increasing the supply voltage (V_{DD}) to boost the gain is not a viable solution, given the breakdown field of the materials. In fact, the scaling of electronic circuits typically involves a reduction in the supply voltage to ensure that the electric field in scaled devices remains below the breakdown field of the material [15]. In the following subsection, some nonlinear blocks, such as a frequency doubler and a mixer, will be analyzed.

4.1.2. Transistor Integrated to Perform Analog Operations

Based on the RF performance of the transistors presented in the previous sections, both an RF mixer and an RF frequency doubler were implemented to further validate their suitability for GHz-range signal processing. To demonstrate the mixer functionality, the device was biased with a drain-source voltage (V_{DS}) of 4 V. Two RF signals, 3.2 GHz and 3.0 GHz, each with a power level of 0 dBm, were combined and applied to the gate of the transistor using a bias-tee, which enabled the superposition of RF and DC signals while preventing direct interaction between the DC and RF sources. A DC gate voltage of 2.5 V was applied, corresponding to the Dirac point of the device.

The output spectrum exhibited the expected frequency components resulting from the mixing of the two input signals. Specifically, signals were observed at 6.0 GHz (−76.3 dBm), 6.2 GHz (−67.2 dBm), and 6.4 GHz (−68.35 dBm), confirming the feasibility of using the proposed device architecture as an RF mixer. The RF power levels were carefully selected to ensure device integrity, considering the sensitivity of the graphene-based active layer.

The frequency doubler was characterized using a similar experimental setup, with the primary distinction being the use of a single RF source and the absence of a signal combiner. The input signal was directly injected into the gate via an RF probe, while the output was collected through a second probe. The device was again biased at V_{DS} = 4 V with a gate voltage of 2.5 V. An input signal of 3 GHz at 0 dBm produced an output component at 6 GHz with a measured power of −72.24 dBm, demonstrating the device's capability to operate as an RF frequency doubler.

4.2. *Performance Assessment of Fabricated Passive Devices*

RF systems are comprised of both active and passive components. After testing and validating active components in the previous sections, the focus now shifts towards passives and the study of the viability of graphene as an enabler for passive RF devices. In recent years, biomedical engineering has experienced remarkable advancements, which were fueled by the integration of emerging materials and wireless technologies. One of those materials is graphene, where, in biosensing, for example, using functionalized graphene has allowed for the development of ultrasensitive biosensors [16–18]. These advancements can revolutionize the field through highly sensitive, real-time detection of biological analytes.

An application of functionalized graphene can be its implementation within antennas. Antennas are necessary for wireless communication, information exchange, and even wireless power transfer. Through the integration of functionalized graphene as the biosensing element in antennas, it should be possible to obtain a wireless biosensor. When functionalized graphene is exposed to a target analyte, that interaction leads to changes in its electrical properties. These changes, in turn, modulate the signal transmitted by the antenna, allowing for the detection and quantification of the target analyte. Through this technique, it should also be possible to combine the benefits of both and obtain a wireless backscatter biosensor. The development of such a system should allow to enable wireless, remote, and non-invasive biosensing with high sensitivity and accuracy.

In this section, the development of a graphene backscatter antenna will be discussed. The design and fabrication of the device will be discussed, as well as the underlying principles and shedding light on the future prospects of such a device.

4.2.1. Wireless Sensing Readout Working Principle

Since graphene is a material with tunable electrical properties that can be controlled by adjusting its chemical potential, it is possible to tune its properties as previously demonstrated and characterized. The interaction of graphene with its surrounding environment, such as target analytes in the case of biosensors, causes its chemical potential to change, thus also changing its impedance. This means that changing the chemical potential of graphene provides valuable information about the interaction between the environmental properties and the graphene surface. By monitoring the impedance change in the graphene biosensor, it is possible to detect and quantify these changes, as is the case with analytes in biosensors.

The tuning of graphene's chemical potential can be achieved through several methods. In the example of biosensors, the most relevant ones are the target analytes coupling with graphene and the voltage gating. Therefore, the change in impedance of graphene biosensors can be effectively modeled or translated by a corresponding change in the gate voltage applied to the graphene layer. These changes in impedance can later be correlated so that a relationship between impedance and gate voltage through experimental calibration can be established, allowing us to interpret the electrical responses of biosensors.

Using RF for wireless power transfer with the objective of powering a biomedical device or sensor and receiving a readout has been the focus of various research works [19–22]. An example is shown in [20], where a solution of wireless power with tracking is presented. It should be possible to implement a biosensor readout by modulating the feedback signal.

In this section, an antenna is implemented with a load that is graphene. To mimic the graphene biosensing capability, graphene will be tuned through voltage gating. Since gating graphene changes its impedance, it should be possible to modify and control the reflected power. Therefore, since graphene is directly connected to the antenna, the amount of reflected power should change according to graphene's impedance. For this experience, the device was fabricated with the surface planarization technique with ion milling.

4.2.2. Sensing Antenna Design

The antenna is similar to a dipole, where two stub-like structures are added on each dipole arm to increase its frequency range. To limit the total size of the antenna, the chosen minimum operation frequency was 3 GHz. To control the resistance of graphene through gating, at least two DC contacts are necessary: a reference and a gate. The antenna is shown on the left side of Figure 17, as well as its dimensions.

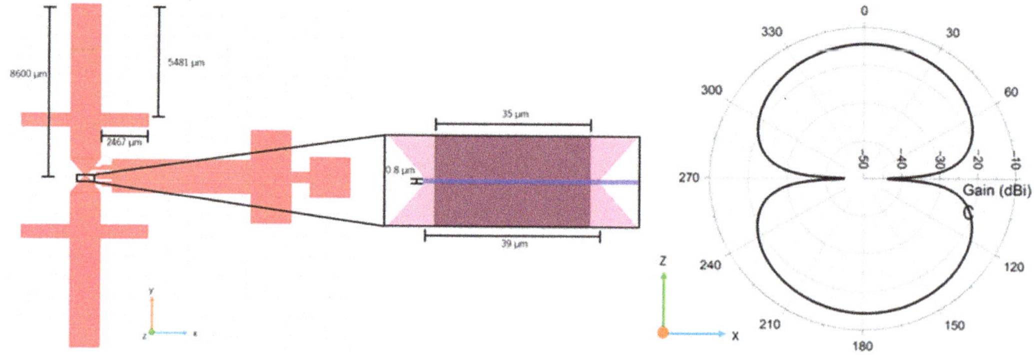

Figure 17. On the left is shown the dipole antenna designed with graphene to obtain a backscatter readout; the inset shows the area with the graphene, as well as its gate and dimensions. On the right is shown the simulated radiation diagram of the antenna.

This antenna was simulated, allowing us to determine its operation frequency at 3 GHz. On the right side of Figure 17, the cross-section of the E-plane radiation diagram at this frequency is shown, where a maximum gain of −14.35 dBi (−11.827 dBi at 6 GHz) can be observed, while its directivity is approximately 3.5 dBi across the simulation frequency range.

4.2.3. Backscattering Antenna Readout

To measure this variation in reflected power, the VNA was calibrated with the Short-Open-Load-Through (SOLT) calibration. To more easily measure the variation in the reflection of the DUT, the load standard was replaced by the antenna. The objective of this modification was to set the reference (the zero reflection) to the background environment. In this way, any change in the graphene load placed at the antenna port will be less difficult to detect. The measurement antenna used was the PE9887-11 horn antenna (Pasternack, Irvine, CA, USA), which has an operating frequency from 4.9 GHz to 7 GHz. The fabricated sample was placed on a wooden table, the horn antenna was placed on top of it, and the DC pads were contacted with needles. The measurement setup block diagram is shown in Figure 18.

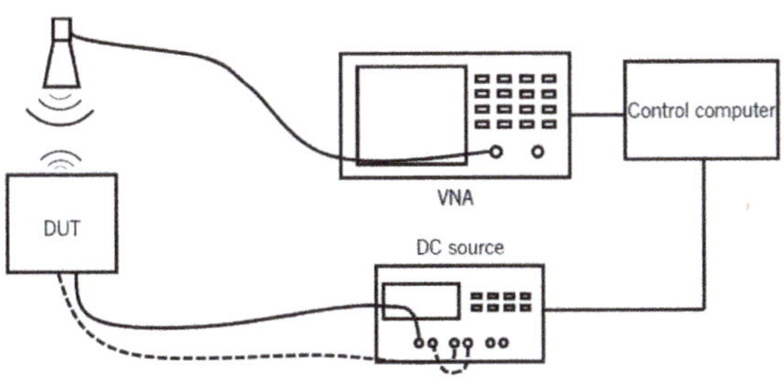

Figure 18. Diagram of the measurement setup used for the measurement of the backscatter of the antenna with graphene on the bottom.

The measurement was performed by changing the gate voltage and measuring the S_{11} through the VNA. The applied voltage ranged from 0 V to 4 V, and the measurement was performed in 1 V steps. The measured values are shown in Figure 19.

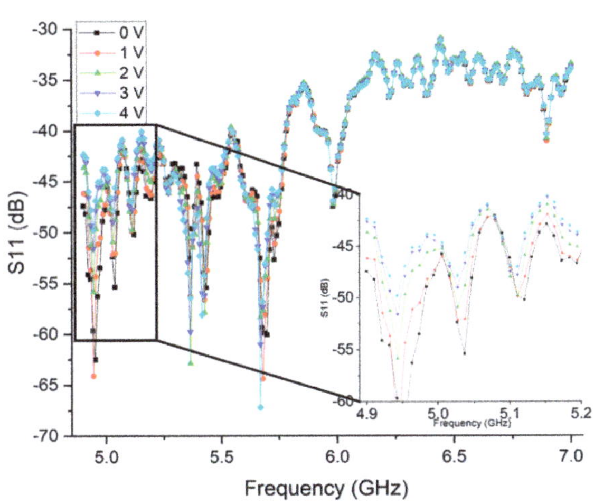

Figure 19. Measured S_{11} in dB of the backscatter by changing the graphene voltage from 0 to 4 volts in the working range of the measurement antenna.

These results show the measured S_{11} for different applied voltages. The inset shown allows us to observe that there is a change in the S_{11} for different applied voltages. At around 4.95 GHz, it is possible to observe that the S_{11} changes from around −65 dB (red curve) up to around −50 dB (light blue curve). The same behavior can be observed for the remaining frequency points, although with a lower variation. These results show that it is possible to control the backscatter of the device by applying a gating voltage to graphene.

5. Conclusions

This work demonstrated the successful development and implementation of two wafer-scale fabrication processes tailored for graphene-based RF devices. The fabricated GFETs exhibited promising RF performance, with intrinsic f_T reaching 14 GHz and fmax up to 80 GHz. The compatibility of the proposed processes with CVD-grown graphene reinforces the scalability of the technology and opens the possibility of extending the methodology to other 2D materials, such as $MoSe_2$. The flexibility of the approach allows for the development of hybrid systems and heterogeneous integration, particularly in applications requiring conformal or biocompatible electronics.

The process flow, device topologies, and layout designs developed in this work provide a viable path forward for integrating graphene RF devices with sensing and actuation functionalities. In particular, the fact that graphene remains exposed throughout much of the fabrication process enables direct functionalization, making the proposed platform highly attractive for biosensing and bio-interfacing applications. An antenna with graphene was developed, enabling passive wireless readout. Through this work, it was demonstrated that the measured S_{11} changed according to an applied gate voltage, allowing it to mimic a graphene sensor. The backscatter readout achieved a maximum variation of around 15 dB.

This study represents a significant step toward enabling reliable, reproducible, and scalable integration of graphene into RF systems. It contributes to the broader effort of transitioning 2D materials from laboratory-scale demonstrations to real-world technologies with impact across communications, sensing, and biomedical domains.

Author Contributions: Conceptualization, V.S., I.C., P.A. and P.M.M.; methodology, V.S., I.C., J.B., P.A. and P.M.M.; software, V.S. and I.C.; validation, V.S., I.C. and J.B.; formal analysis, V.S. and I.C.; investigation, V.S., I.C. and J.B.; resources, P.A. and P.M.M.; data curation, V.S., I.C. and H.D.; writing—original draft preparation, V.S., I.C., H.D. and P.M.M.; writing—review and editing, V.S., I.C., H.D., J.B. and P.M.M.; supervision, J.B., P.A. and P.M.M.; funding acquisition, P.M.M. All authors have read and agreed to the published version of the manuscript.

Funding: This research was funded by the Portuguese Foundation for Science and Technology through grants SFRH/BD/141462/2018, COVID/BD/153002/2022, SFRH/BD/137529/2018, and COVID/BD/153232/2023. This work was also supported by the Portuguese Foundation for Science and Technology under the national support to R&D units grant, with the reference projects UIDB/04436/2020 and UIDP/04436/2020.

Data Availability Statement: The original contributions presented in this study are included in the article. Further inquiries can be directed to the corresponding author.

Conflicts of Interest: The authors declare no conflicts of interest.

References

1. Lee, S.; Zhong, Z. Nanoelectronic Circuits Based on Two-Dimensional Atomic Layer Crystals. *Nanoscale* **2014**, *6*, 13283–13300. [CrossRef] [PubMed]
2. Zhu, W.; Perebeinos, V.; Freitag, M.; Avouris, P. Carrier Scattering, Mobilities, and Electrostatic Potential in Monolayer, Bilayer, and Trilayer Graphene. *Phys. Rev. B Condens. Matter Mater Phys.* **2009**, *80*, 235402. [CrossRef]

3. Wu, Y.Q.; Farmer, D.B.; Valdes-Garcia, A.; Zhu, W.J.; Jenkins, K.; Dimitrakopoulos, C.; Avouris, P.; Lin, Y.-M. Record High RF Performance for Epitaxial Graphene Transistors. In Proceedings of the International Electron Devices Meeting (IEDM), Washington, DC, USA, 5–7 December 2011; Volume 11, pp. 3690–3693.
4. Peng, M.; Cheng, J.; Zheng, X.; Ma, J.; Feng, Z.; Sun, X. 2D-Materials-Integrated Optoelectromechanics: Recent Progress and Future Perspectives. *Rep. Prog. Phys.* **2023**, *86*, 026402. [CrossRef] [PubMed]
5. Xue, H.; Gao, W.; Gao, J.; Schneider, G.F.; Wang, C.; Fu, W. Radiofrequency Sensing Systems Based on Emerging Two-Dimensional Materials and Devices. *Int. J. Extrem. Manuf.* **2023**, *5*, 032010. [CrossRef]
6. Colmiais, I.; Silva, V.; Borme, J.; Alpuim, P.; Mendes, P.M. Towards RF Graphene Devices: A Review. *FlatChem* **2022**, *35*, 100409. [CrossRef]
7. Cheng, R.; Bai, J.; Liao, L.; Zhou, H.; Chen, Y.; Liu, L.; Lin, Y.C.; Jiang, S.; Huang, Y.; Duan, X. High-Frequency Self-Aligned Graphene Transistors with Transferred Gate Stacks. *Proc. Natl. Acad. Sci. USA* **2012**, *109*, 11588–11592. [CrossRef] [PubMed]
8. Gao, Q.; Xiong, X.; Li, X.; Wu, Y. High Performance Double-Gate Graphene Radio-Frequency Transistors. In Proceedings of the 2018 IEEE International Conference on Electron Devices and Solid State Circuits (EDSSC), Shenzhen, China, 6–8 June 2018.
9. Badmaev, A.; Che, Y.; Li, Z.; Wang, C.; Zhou, C. Self-Aligned Fabrication of Graphene Rf Transistors with t-Shaped Gate. *ACS Nano* **2012**, *6*, 3371–3376. [CrossRef] [PubMed]
10. Liao, C.-D.; Capasso, A.; Queirós, T.; Domingues, T.; Cerqueira, F.; Nicoara, N.; Borme, J.; Freitas, P.; Alpuim, P. Optimizing PMMA Solutions to Suppress Contamination in the Transfer of CVD Graphene for Batch Production. *Beilstein J. Nanotechnol.* **2022**, *13*, 796–806. [CrossRef] [PubMed]
11. Park, D.-W.; Mikael, S.; Chang, T.-H.; Gong, S.; Ma, Z. Bottom-Gate Coplanar Graphene Transistors with Enhanced Graphene Adhesion on Atomic Layer Deposition Al_2O_3. *Appl. Phys. Lett.* **2015**, *106*, 102106. [CrossRef]
12. Heydrich, S.; Hirmer, M.; Preis, C.; Korn, T.; Eroms, J.; Weiss, D.; Schuller, C. Scanning Raman Spectroscopy of Graphene Antidot Lattices: Evidence for Systematic p-Type Doping. *Appl. Phys. Lett.* **2010**, *97*, 43113. [CrossRef]
13. Levesque, P.L.; Sabri, S.S.; Aguirre, C.M.; Guillemette, J.; Siaj, M.; Desjardins, P.; Szkopek, T.; Martel, R. Probing Charge Transfer at Surfaces Using Graphene Transistors. *Nano Lett.* **2011**, *11*, 132–137. [CrossRef] [PubMed]
14. Vieira, N.C.S.; Borme, J.; Machado, G.; Cerqueira, F.; Freitas, P.P.; Zucolotto, V.; Peres, N.M.R.; Alpuim, P. Graphene Field-Effect Transistor Array with Integrated Electrolytic Gates Scaled to 200 Mm. *J. Phys. Condens. Matter* **2016**, *28*, 085302. [CrossRef] [PubMed]
15. Bianchi, M.; Guerriero, E.; Fiocco, M.; Alberti, R.; Polloni, L.; Behnam, A.; Carrion, E.A.; Pop, E.; Sordan, R. Scaling of Graphene Integrated Circuits. *Nanoscale* **2015**, *7*, 8076–8083. [CrossRef] [PubMed]
16. Cabral, P.D.; Domingues, T.; Machado, G.; Chicharo, A.; Cerqueira, F.; Fernandes, E.; Athayde, E.; Alpuim, P.; Borme, J. Clean-Room Lithographical Processes for the Fabrication of Graphene Biosensors. *Materials* **2020**, *13*, 5728. [CrossRef] [PubMed]
17. Abrantes, M.; Rodrigues, D.; Domingues, T.; Nemala, S.S.; Monteiro, P.; Borme, J.; Alpuim, P.; Jacinto, L. Ultrasensitive Dopamine Detection with Graphene Aptasensor Multitransistor Arrays. *J. Nanobiotechnol.* **2022**, *20*, 495. [CrossRef] [PubMed]
18. Palacio, I.; Moreno, M.; Náñez, A.; Purwidyantri, A.; Domingues, T.; Cabral, P.D.; Borme, J.; Marciello, M.; Mendieta-Moreno, J.I.; Torres-Vázquez, B.; et al. Attomolar Detection of Hepatitis C Virus Core Protein Powered by Molecular Antenna-like Effect in a Graphene Field-Effect Aptasensor. *Biosens. Bioelectron.* **2023**, *222*, 115006. [CrossRef] [PubMed]
19. Dinis, H.; Colmiais, I.; Mendes, P.M. A Multiantenna Approach to Maximize Wireless Power Transfered to Implantable Devices. In Proceedings of the 2017 International Applied Computational Electromagnetics Society Symposium-Italy (ACES), Firenze, Italy, 26–30 March 2017. [CrossRef]
20. Colmiais, I.; Dinis, H.; Mendes, P.M. WPT System for Implantable Devices Using a Phased Array and Tracking Algorithm for Freely Moving Rats. In Proceedings of the 2019 IEEE 6th Portuguese Meeting on Bioengineering (ENBENG), Lisbon, Portugal, 22–23 February 2019; IEEE: Los Alamitos, CA, USA, 2019; pp. 1–4.
21. Dinis, H.; Colmiais, I.; Mendes, P.M. Extending the Limits of Wireless Power Transfer to Miniaturized Implantable Electronic Devices. *Micromachines* **2017**, *8*, 359. [CrossRef] [PubMed]
22. Dinis, H.; Colmiais, I.; Mendes, P.M. Design and Validation of a Six-Antenna WPT System with Tracking Capabilites for Biomedical Devices. In Proceedings of the 2019 17th IEEE International New Circuits and Systems Conference (NEWCAS), Munich, Germany, 23–26 June 2019; pp. 1–4. [CrossRef]

nanomaterials

MDPI

Article

Optimizing the Structure and Performances of Cu-MOF@$Ti_3C_2T_X$ Hybrid Electrodes by Introducing Modulated Ligand

Sumin Li [1,*], Xiaokun Qu [1], Feng Liu [1,2], Pingwei Ye [2,*], Bo Yang [2], Qiang Cheng [1], Mengkun Yang [1], Yijing Nie [1] and Maiyong Zhu [1]

1 School of Materials Science & Engineering, Jiangsu University, Zhenjiang 212013, China; 2212205064@stmail.ujs.edu.cn (X.Q.); 2222205025@stmail.ujs.edu.cn (F.L.); cq18552374534@163.com (Q.C.); 2222305060@stmail.ujs.edu.cn (M.Y.); nieyijing@ujs.edu.cn (Y.N.); maiyongzhu@ujs.edu.cn (M.Z.)

2 State Key Laboratory of NBC Protection for Civilian, Beijing 102205, China; 12426035@zju.edu.cn

* Correspondence: li_sm@ujs.edu.cn (S.L.); yepw2001@163.com (P.Y.)

Abstract: To date, two-dimensional metal–organic frameworks (2D MOFs) have attracted much attention in many fields. Owing to their ultra-high porosity and specific surface area, great structural diversity and functional tunability, as well as feasible precision design at the molecular level, 2D MOFs have won rapid development in the field of energy storage. However, as a coordination compound, MOFs possess poor structural stability and are prone to structural collapse in electrochemical reactions, which seriously limits their electrochemical performance. Therefore, there is an urgent need to improve the structural stability of MOF electrode materials. In this study, a 2D MOF@$Ti_3C_2T_X$ hybrid was constructed, in which urea pyrimidinone isocyanate (UPy-NCO) units were introduced via a condensation reaction with the active functional groups on MOFs, thus forming multiple hydrogen bonds among MOF frameworks to strengthen their structural stability. Importantly, 2,6-diaminopyridine was utilized to modulate the structure and properties. Initially, the mono-coordination model of the N atom on a pyridine ring with metal ions could create defects and form further pores. Two $-NH_2$ groups helped to improve the grafting reaction degree of UPy-NCO, leading to an increased ratio of forming quadruple hydrogen bonds (H-bonds), further strengthening the structure of the hybrid. As expected, the Cu-MOF@$Ti_3C_2T_X$-20%DAP-UPy hybrid exhibited a specific capacitance of 148 F g^{-1} at 1 A g^{-1}, which is 45% higher than that of Cu-MOF@$Ti_3C_2T_X$-UPy (102 F g^{-1}). A good capacitance retention of 88% was obtained as the current density increased from 0.2 to 5 A g^{-1}. Moreover, excellent cycling stability (91.1%) was obtained at 1 A g^{-1} after 5000 cycles.

Keywords: 2D MOF; multiple hydrogen bonds; hybrid; supercapacitors

1. Introduction

With the rapid development of portable electronics, supercapacitors have attracted much attention due to their fast charging and discharging, high power density, green environmental protection, and long lifetime [1–6]. Due to their ultra-high porosity, large specific surface area, and structural tunability [7–15], MOFs have received widespread attention in supercapacitor and other energy-storage applications [16–19]. In particular, owing to their exposed active sites and higher electrical conductivity, as well as regular and well-ordered pore structure [20–27], 2D MOFs are more conducive to achieving high performance as electrode materials. However, the weak coordination bonds between metal ions and organic ligands make MOF frameworks unstable and even prone to collapse, leading to poor

https://doi.org/10.3390/nano15110864

cycling stability [28–34], thus restricting their practical applications. Therefore, enhancing the cycling stability of MOF energy-storage materials needs to be addressed urgently.

Self-healing ability provides a novel design idea for improving the service life of energy storage devices. Currently, two self-healing categories are mainly investigated: externally assisted self-healing and intrinsic self-healing [35–37]. Externally assisted self-healing is realized by incorporating self-healing agents into the materials [38–40]. For example, Guo [41] et al. investigated epoxy matrix microencapsulated self-healing materials. However, most microcapsule-based self-repairs exhibit only onetime self-repair properties, which severely limits their practical applications. Intrinsic self-healing materials can repair by themselves through the introduction of reversible chemical bonds [37,41–45] (e.g., boronic acid diester bonds, hydrogen bonds, etc.). For example, in 2023, Wang [45] et al. sprayed hydrophobically modified CNT and NiO/CoO nanoparticles onto the surface of a hydrogel (PVA) to construct electrodes, which utilized the dynamic hydrogen bonds formed between PVA and HCl through physical cross-linking to provide the self-healing ability. The as-prepared supercapacitor achieved a capacitance retention of 75.4% after 2000 cycles. Both the introduction of self-healing agents and reversible chemical bonds can give energy storage devices self-healing abilities, which is important for prolonging their service life [46–51]. However, traditional self-healing materials are usually polymers, which are not electrochemically active and thus unfavorable for producing capacitance, leading to the decreased electrochemical performance of electrodes or energy storage devices [52–58].

We have proposed an innovative strategy to enhance the structural stability of MOFs by introducing quadruple hydrogen bonds directly into MOF frameworks [23]. Here, to further enhance the structural stability and performance of MOF-based electrodes, a modulated linker, 2,6-diaminopyridine, was utilized. On the one hand, the mono-coordination model of the N atom on a pyridine ring with metal ions could produce defects and form further pores. On the other hand, the two $-NH_2$ groups help to improve the grafting reaction degree of UPy-NCO units, improving the ratio of forming quadruple hydrogen bonds, further strengthening their structure stability.

2. Materials and Methods

2.1. Preparation of Samples

The materials, reagents, and synthesis of $Ti_3C_2T_X$, copper oxide nanosheets (Cu-ONS), Cu-MOF@$Ti_3C_2T_X$, and UPy-NCO are displayed in the Supporting Information (SI).

Synthesis of Cu-MOF@$Ti_3C_2T_X$-DAP: H_2BDC-NH_2 (30 mg), 2,6-diaminopyridine (0, 3, 6, 9 mg), and PVP (50 mg) were dissolved in a mixed solvent containing 1.1 mL of DMF, 1.1 mL of deionized (DI) water, and 1.1 mL of ethanol, obtaining a mixed solution. Cu-ONS (25 mg) was dispersed in 5 mL of DI water under stirring, and then $Ti_3C_2T_X$ (10 mg) was added, forming another solution. Then, the two solutions were together transferred to a reactor and remained at 100 °C for 20 h. Lastly, the sample was washed and dried, and denoted as Cu-MOF@$Ti_3C_2T_X$-X%DAP (X = 10, 20 and 30).

Synthesis of Cu-MOF@$Ti_3C_2T_X$-DAP-UPy: 40 mg of Cu-MOF@$Ti_3C_2T_X$-DAP was dissolved in 20 mL of DMF and 30 mg of UPy-NCO was dissolved in 10 mL of DMF. Then, the two solutions were mixed, followed by the addition of 2 drops of dibutyltin dilaurate. The mixed solution was maintained at 90 °C for 20 h under stirring. Finally, the sample was washed with ethanol, dried at 40 °C, and denoted as Cu-MOF@$Ti_3C_2T_X$-DAP-UPy, as described in Scheme 1.

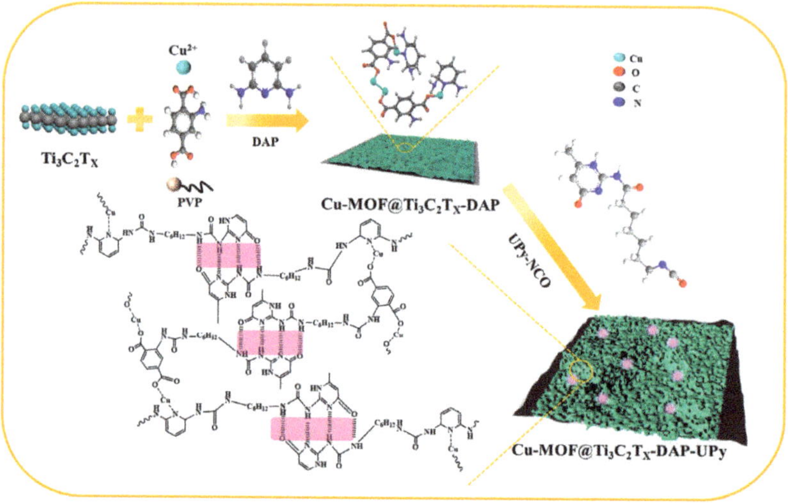

Scheme 1. Schematic illustration of Cu-MOF@$Ti_3C_2T_X$-DAP-UPy.

2.2. Electrochemical Characterization

The working electrode material consisted of acetylene black, polyvinylidene chloride (PVDF) and active material. The three materials were mixed (mass ratio of 8:1:1) with N-methylpyrrolidone to form a homogeneous paste. Then, the paste was coated on a piece of Ni foam (1 × 1 cm^2) current collector, followed by a drying process at 80 °C for 12 h, thus obtaining a working electrode (the loading amount was 1.5~2 mg cm^{-2}).

An electrochemical workstation (CHI760E) was used to perform electrochemical experiments including cyclic voltammetry (CV), constant current charge–discharge (GCD) and electrochemical impedance spectroscopy (EIS). A three-electrode system was used to investigate the performance of single electrodes, in which the working electrode, counter electrode, and reference electrode were the as-prepared material, Pt plate, and Hg/HgO, respectively. A 1 M KOH aqueous solution was used as the electrolyte. The capacitance (C, F g^{-1}) values were evaluated by calculating the integral of the area under the charging and discharging part of the curves as follows [48]:

$$C = I\int \left(1/m \times V(t)\right)dt = I\Delta t/m\Delta V \tag{1}$$

where I (A g^{-1}) is the current density, Δt (s) is the discharge time, m (g) is the mass of the active material, and ΔV (V) is the potential.

In addition, the energy density (E, Wh kg^{-1}), power density (P, W kg^{-1}) and Coulombic Efficiency (CE) of the prepared device was investigated using the following formulae [56]:

$$E = \frac{C \times \Delta V^2}{2 \times 3.6} \tag{2}$$

$$P = \frac{3600 \times \mathrm{E}}{\Delta \mathrm{t}} \tag{3}$$

$$CE = \frac{I_d t_d}{I_c t_c} \times 100\% \tag{4}$$

where Δt (s) is the discharge time, ΔV (V) is the potential window, I_d (A) is the discharge current, t_d (s) is the discharge time, I_c (A) is the charge current, and t_c (s) is the charge time.

3. Results

3.1. Structural Characterization

The SEM images of the samples are shown in Figure 1 and Figure S1. As displayed in Figure S1a, lots of nanosheets with a size of about 1 μm were observed in Cu-MOF@$Ti_3C_2T_X$. With the addition of 2,6-diaminopyridine, the microstructure of the samples changed obviously. The size of the lamellar structure became smaller with the increasing 2,6-diaminopyridine content. The mono-coordination model of the N atom on the pyridine ring with metal ions created defects and formed further pores, leading to the decrease in size of the nanosheets (Figure S1b,c). For Cu-MOF@$Ti_3C_2T_X$-30%DAP, lots of irregular fragments can be seen (Figure S1d), caused by the large number of defects in the nanosheets due to the excess of 2,6-diaminopyridine [59]. After the UPy-NCO units were grafted via a condensation reaction with the $-NH_2$ active groups on the MOFs, the morphology of the Cu-MOF@$Ti_3C_2T_X$-DAP-UPy samples (Figure 1) were further changed. Especially, the Cu-MOF@$Ti_3C_2T_X$-30%DAP-UPy (Figure 1c) exhibited a completely different morphology compared with Cu-MOF@$Ti_3C_2T_X$-30%DAP (Figure S1d), showing a thinner layer and cross-linked network structure. As the amount of 2,6-diaminopyridine increased up to 30%, more $-NH_2$ active groups were introduced onto the MOFs, leading to an improvement in the grafting reaction degree of UPy-NCO and a further increase in the ratio of forming quadruple hydrogen bonding, subsequently promoting the formation of a cross-linked network structure. The EDS elemental analysis (Figure 1d) shows that the elements, N, O, Cu, and Ti, are uniformly distributed in Cu-MOF@$Ti_3C_2T_X$-20%DAP-UPy sample.

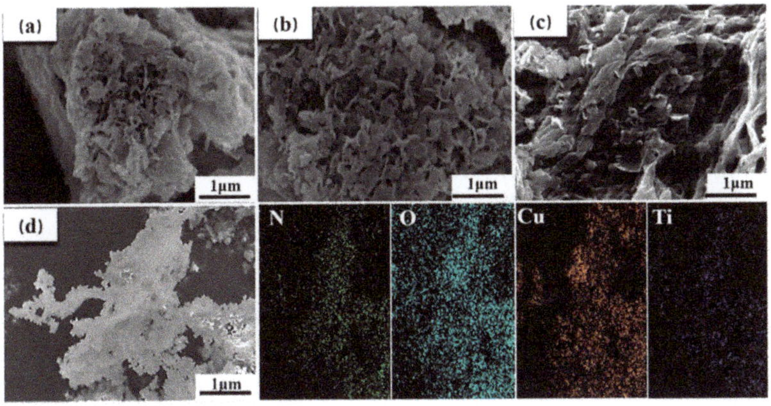

Figure 1. SEM images of the samples. (**a**) Cu-MOF@$Ti_3C_2T_X$-10%DAP-UPy, (**b**) Cu-MOF@$Ti_3C_2T_X$-20%DAP-UPy, (**c**) Cu-MOF@$Ti_3C_2T_X$-30%DAP-UPy, and (**d**) EDS images of Cu-MOF@$Ti_3C_2T_X$-20%DAP-UPy with the corresponding elemental distributions of N, O, Cu and Ti.

The N_2 adsorption–desorption tests were performed to investigate the pore characteristics of the prepared materials, as shown in Figure 2. All the samples exhibit type IV curves (Figure 2a), indicating the co-existence of micropores and mesopores [60]. The hierarchical pores, including micropores (1.5 nm), mesopores (8, 17, 37 and 50 nm), and macropores (50–86 nm), are clearly displayed in Figure 2b,c. Most importantly, an 8 nm mesopore can be noticed in both Cu-MOF@$Ti_3C_2T_X$-20%DAP and Cu-MOF@$Ti_3C_2T_X$-20%DAP-UPy, which might be due to the defect originating from the mono-coordination model between 2,6-diaminopyridine and Cu^{2+}. Moreover, the specific surface areas of Cu-MOF@$Ti_3C_2T_X$, Cu-MOF@$Ti_3C_2T_X$-20%DAP, and Cu-MOF@$Ti_3C_2T_X$-20%DAP-UPy are 43.1, 27.4, and 14.6 $m^2\ g^{-1}$, respectively. As a modulated linker, 2,6-diaminopyridine participated in the growth of MOFs, as well as the grafting reactions with UPy-NCO units, resulting in a decrease in the proportion of 1.6 nm micropores, thus causing a decrease in the specific surface area.

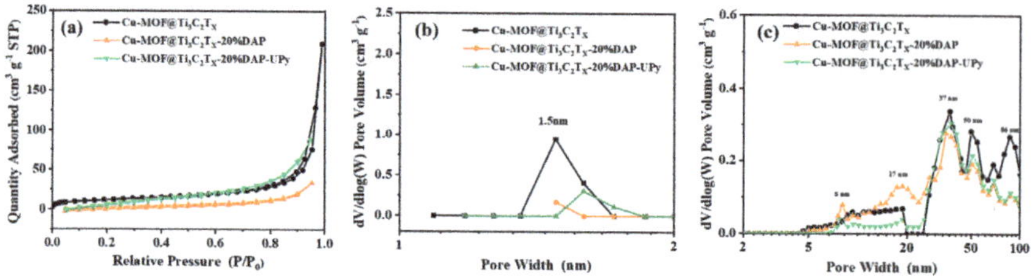

Figure 2. (**a**) N_2 adsorption–desorption isotherms, (**b**) DFT pore-size distribution curves and (**c**) BJH pore-size distribution curves of the samples.

The XRD patterns of the samples are displayed in Figure S2. Compared with Cu-MOF@$Ti_3C_2T_X$-UPy, the samples containing 2,6-diaminopyridine exhibit new diffraction peaks at 2θ of 5.9°, 11.9°, 25.3°, 26.8°, and 43.2°, which is due to the formation of new crystal structures stemming from the coordinate reactions between Cu^{2+} and N atoms on the pyridine rings [61].

Figure 3a shows the FT-IR spectra of the samples. For Cu-MOF@$Ti_3C_2T_X$, the two peaks at 3237 and 3142 cm^{-1} are ascribed to the $-NH_2$ stretching vibrations [62], while the strong and sharp characteristic peak at 1689 cm^{-1} is caused by the C=O stretching vibrations. For the samples grafted with UPy-NCO, several obvious changes can be noticed in the peaks. First, the broad peak at 3454 cm^{-1} associated with the −OH is strengthened in intensity, which is attributed to the formation of intermolecular H-bonds (C=O···H−N). Second, the characteristic peaks at 2936 and 2861 cm^{-1} can be noticed, which are caused by the $-CH_2$ stretching vibrations, implying that the UPy-NCO units were successfully grafted onto the MOFs. Most importantly, the characteristic peak corresponding to the C=O stretching vibrations exhibited a redshift in wavenumber and presented a blunt shape in profile, which is also associated with the intermolecular H−bonds (C=O···H−N). As the content of 2,6-diaminopyridine increased, the characteristic peak of C=O gradually became broader, proving that more hydrogen bonds were formed. Particularly, for Cu-MOF@$Ti_3C_2T_X$-30%DAP-UPy a broad adsorption band at around 1640–1380 cm^{-1} was formed, caused by the peak shift of C=O, as well as the peak overlap of different groups including C=C and C=N. It is worth stressing that the −NH bonds are also an important component taking part in the formation of quadruple hydrogen bonds, which can be further identified by the variable FT-IR spectra (Figure 3b). When the temperature was increased from 25 °C to 125 °C, with the dissociation of hydrogen bonds, the peak associated with the stretching vibration of the −NH bonds shifted slightly from 3147.8 cm^{-1} to 3152.7 cm^{-1} [63,64].

Figure 4 shows the XPS spectra of Cu-MOF@$Ti_3C_2T_X$ and Cu-MOF@$Ti_3C_2T_X$-20%DAP-UPy. As displayed in Figure 4a, with the introduction of 2,6-diaminopyridine, the peak intensity of the N element in Cu-MOF@$Ti_3C_2T_X$-20%DAP-UPy is highlighted compared to that of Cu-MOF@$Ti_3C_2T_X$, which is attributed to the increase in grafting density of the UPy-NCO units. Moreover, in the N 1s spectra of Cu-MOF@$Ti_3C_2T_X$-20%DAP-UPy (Figure 4b), the peak corresponding to the Cu−N (399.3–399.7 eV) can be observed, further confirming the coordinate reaction between the N atoms on the pyridine rings and Cu^{2+} ions. The other two peaks are assigned to C−N on pyridine (400.3–400.7 eV), and C−N formed on the amide bond (398.3–399 eV) [65].

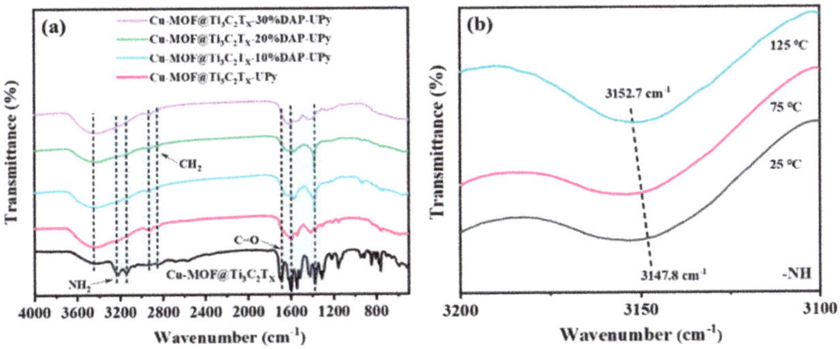

Figure 3. (**a**) FT−IR spectra of samples and (**b**) temperature-variable FT-IR spectra of Cu-MOF@$Ti_3C_2T_X$-20%DAP-UPy.

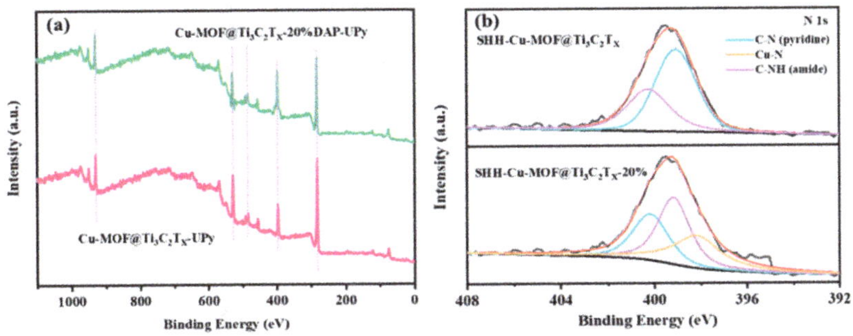

Figure 4. XPS spectra of samples: (**a**) survey spectrum, (**b**) high-resolution XPS spectra of N 1s.

3.2. Electrochemical Performance

The electrochemical properties of the prepared samples are shown in Figure 5. The CV curves of different samples at a scan rate of 50 mV s^{-1} are shown in (Figure 5a). Obvious redox peaks can be observed in CV curves, which stems from the redox reactions of Cu−O clusters in the Cu-MOF as follows:

$$Cu{-}O + OH^- \leftrightarrow CuOOH + e^- \quad (5)$$

Moreover, the introduction of $Ti_3C_2T_X$ could also produce pseudo-capacitance [66,67]. The combination with non-ideal triangles in the GCD curves (Figure 5b) suggests the two electrochemical behaviors coexist: pseudo-capacitance and double-layer capacitance. Moreover, the longer discharging times indicate that the samples with the addition of 2,6-diaminopyridine exhibit higher capacitances compared with Cu-MOF@$Ti_3C_2T_X$-UPy. As mentioned above, the mono-coordination model of the N atom on the pyridine ring with Cu^{2+} ions could create defects and form further pores, exposing more active sites and facilitating redox reactions, leading to enhanced capacitances. However, abundant defects and pores prolong the ion transportation path and reduce the ion diffusion rate, which can be demonstrated by the lower slope of the straight lines in the low-frequency region (Figure 5c). Notably, the Cu-MOF@$Ti_3C_2T_X$-20%DAP-UPy exhibits the best electrochemical performance, delivering a specific capacitance of 148 F g^{-1} at 1 A g^{-1}. With the increased scan rates (Figure 5d), similar shapes of CV curves were maintained, indicating good reversibility.

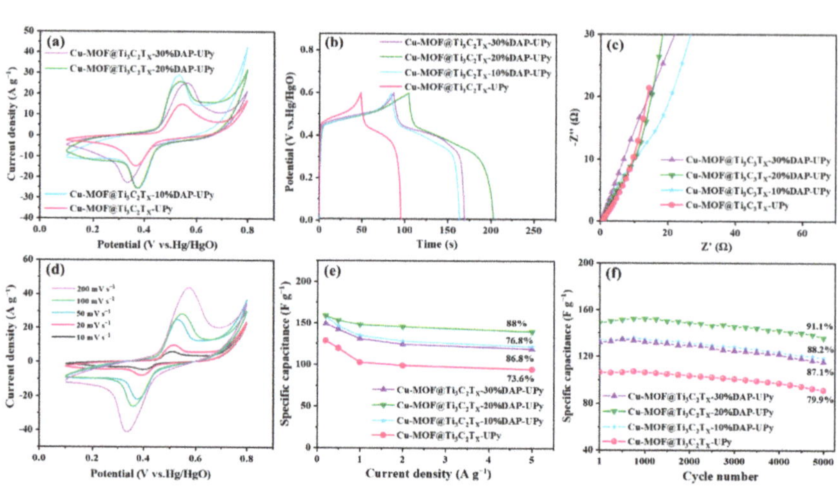

Figure 5. Electrochemical performances of samples: (**a**) CV profiles at 50 mV s^{-1}, (**b**) GCD profiles at 1 A g^{-1}, (**c**) Nyquist plots, (**d**) CV profiles of Cu-MOF@$Ti_3C_2T_X$-20%DAP-UPy, (**e**) specific capacitances at various current densities, and (**f**) cycling performances at 1 A g^{-1}.

The calculated capacitances investigating the effect of 2,6-diaminopyridine on the performances of the samples are shown in Figure 5e. With the current density increased from 0.2 to 5 A g^{-1}, the capacitance retentions of Cu-MOF@$Ti_3C_2T_X$-UPy, Cu-MOF@$Ti_3C_2T_X$-10%DAP-UPy, Cu-MOF@$Ti_3C_2T_X$-20%DAP-UPy, and Cu-MOF@$Ti_3C_2T_X$-30%DAP-UPy are 73.6%, 76.8%, 88%, and 86.8%, respectively. Furthermore, the cycling performances were investigated, as shown in Figure 5f. After 5000 cycles, the capacitance retentions of the above samples are 79.9%, 88.2%, 91.1% and 87.1%, respectively. The participation of 2,6-diaminopyridine not only improves the storage capacitance of the samples, but also optimizes their rate performance and cycle stability. The $-NH_2$ group on 2,6-diaminopyridine helps to facilitate the grafting reaction of UPy-NCO, and promotes the formation of quadruple hydrogen bonding, thus strengthening the structural stability of the hybrid. However, excessive 2,6-diaminopyridine accordingly produces more defects, weakening the crystal structure of the MOF; thus, the Cu-MOF@$Ti_3C_2T_X$-30%DAP-UPy shows a decrease in the rate and cycle performances. As shown in Figure 6, compared with the other two samples, the Cu-MOF@$Ti_3C_2T_X$-20%DAP-UPy presents a relatively complete surface structure. The Cu-MOF@$Ti_3C_2T_X$-20%DAP-UPy exhibits a high capacitance retention of 91.1% after 5000 cycles, which is superior to some other MOF-based electrodes in the literature (e.g., Table 1).

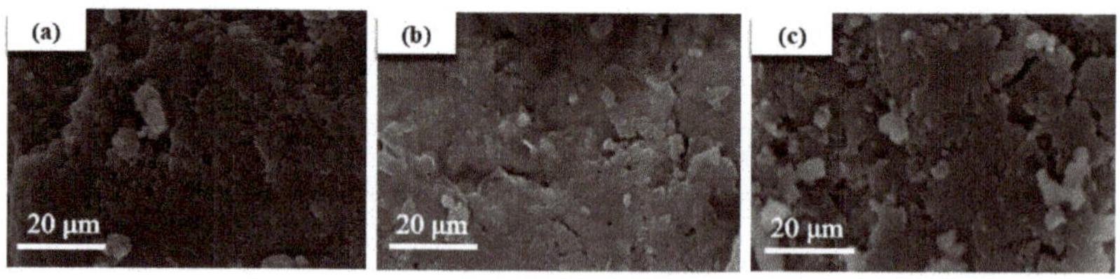

Figure 6. SEM images of (**a**) Cu-MOF@$Ti_3C_2T_X$-10%DAP-UPy, (**b**) Cu-MOF@$Ti_3C_2T_X$-20%DAP-UPy, and (**c**) Cu-MOF@$Ti_3C_2T_X$-30%DAP-UPy electrodes after cycling tests.

Table 1. Cycle stability of MOF-based electrode materials.

Electrode Materials	Electrolyte	Cyclic Stability	Ref.
$Ni_3(HITP)_2$	0.5 M Na_2SO_4	84% (after 10,000 cycles 0.1 mA cm^{-2})	[68]
Cu-CAT NWAs	3 M KCl	80% (after 5000 cycles 0.5 A g^{-1})	[69]
Ni/Co-BTC (2:1)	6 M KOH	60% (after 3000 cycles 5.0 A g^{-1})	[70]
NiCo-MOF	2 M KOH	76.3% (after 5000 cycles 5.0 A g^{-1})	[71]
Ni@Cu-MOF	6 M KOH	63% (after 5000 cycles 1.0 A g^{-1})	[72]
Cu-MOF/rGO	1 M Na_2SO_4	83.3% (after 4000 cycles 10 A g^{-1})	[73]
Cu-MOF@$Ti_3C_2T_X$-20%DAP-UPy	1 M KOH	91.1% (after 5000 cycles 1.0 A g^{-1})	This work

The kinetic mechanism of the as-obtained hybrid was investigated by using the b-value model as follows [69]:

$$i = a\nu^{b} \tag{6}$$

where i is the peak current and v is the scan rate. Especially, b is an adjustable parameter and has two critical values of 0.5 and 1.0, referring to the diffusive-controlled behavior and surface capacitive-controlled behavior, respectively.

In addition, to reveal the energy-storage mechanism in detail the Dunn's model was also used, in which the contributions of diffusive-controlled behavior and surface capacitive-controlled behavior are descripted as follows [70]:

$$i(V) = k_1\nu + k_2\nu^{0.5} \tag{7}$$

where $k_1\nu$ and $k_2\nu^{0.5}$ correspond to the capacitive and diffusion-controlled currents, respectively. As shown in Figure 7a, for the Cu-MOF@$Ti_3C_2T_X$-20%DAP-UPy, the calculated b values of the oxidation and reduction peaks are 0.66 and 0.74, respectively, indicating that the charge storage process includes both diffusive-controlled behavior and surface capacitive behavior. Combining the calculated contributions of the two charge storage behaviors (Figure 7b,c), it is obvious that the diffusive contribution is superior to the surface capacitive contribution even at 50 mV s^{-1}. The formation of defects derived from the coordination reaction between 2,6-diaminopyridine and Cu^{2+} ions could promote ions to thoroughly penetrate materials, making full use of active sites to produce more capacitance. With the increased scan rates, the ions cannot diffuse sufficiently inside active materials in time, thus the diffusive-controlled proportion decreased while the surface capacitive contribution increased.

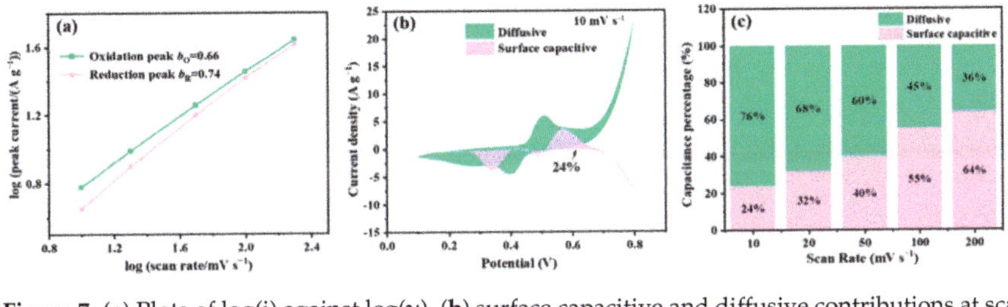

Figure 7. (**a**) Plots of log(i) against log(ν), (**b**) surface capacitive and diffusive contributions at scan rate of 10 mV s^{-1}, and (**c**) capacitance contribution ratios.

3.3. Electrochemical Properties of ASC Device

An asymmetric supercapacitor (Cu-MOF@$Ti_3C_2T_X$-20%DAP-UPy//AC) was assembled using Cu-MOF@$Ti_3C_2T_X$-20%DAP-UPy as the positive electrode, activated carbon (AC) (Figure S3) as the negative electrode, and 1 M KOH aqueous solution as the electrolyte, respectively.

The CV curves of the device under different voltage windows from 0–1.0 to 0–1.8 V are displayed in Figure 8a. When the voltage window was set as 1.7–1.8 V, a slight polarization phenomenon can be observed, thus the 0–6 V was chosen as the appropriate working voltage. The CV curves at different scan rates are shown in Figure 8b, exhibiting similar shapes without obvious deformation, indicating that the device possesses good capacitive behavior even under fast charging/discharging. Based on the GCD curves of the device (Figure 8c), the specific capacitances were calculated and are displayed in Figure 8d. At 0.5 A g^{-1}, the device delivers a specific capacitance of 50 F g^{-1}, exhibiting a capacitance retention of 85% as the current density was increased to 5 A g^{-1}. A maximum energy density of 17.9 W h kg^{-1} was obtained (Figure 8e), which is superior to that of some reported MOF-based devices [64,74–78]. The cycle stability of the device is shown in Figure 8f: a capacitance retention of 85% was delivered after 5000 cycles at 2 A g^{-1}. Moreover, the calculated coulombic efficiency was 90%.

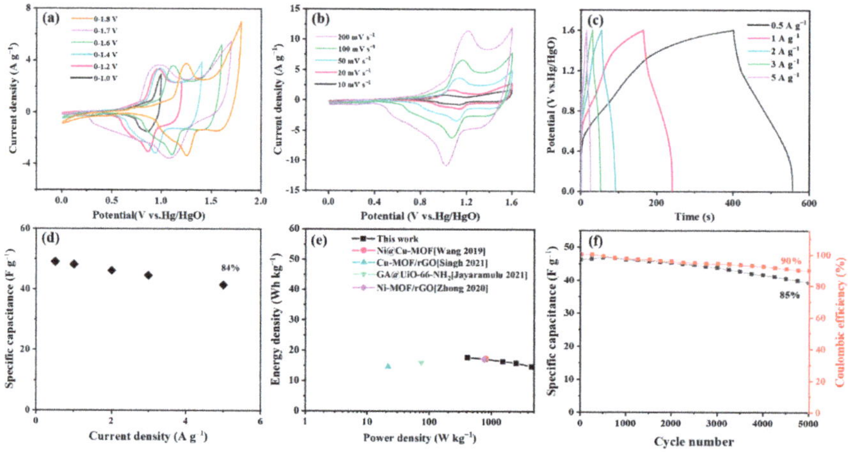

Figure 8. Electrochemical performance of Cu-MOF@$Ti_3C_2T_X$-20%DAP-UPy//AC device. (**a**) CV profiles at various voltage windows at scan rate of 50 mV s^{-1}, (**b**) CV profiles at various scan rates, (c) GCD curves under different current densities, (**d**) specific capacitances at different current densities, (**e**) Ragone plots [72–75], (**f**) cycling stability and coulombic efficiency at current density of 2 A g^{-1}.

4. Discussion

To enhance the structural stability of MOF-based electrodes, we have explored a strategy of introducing quadruple hydrogen bonds onto MOFs. Here, to further promote the introduction of quadruple hydrogen bonds and optimize the performance of 2D MOF electrodes, 2,6-diaminopyridine was used as a modulated linker to participate in the growth of MOF and the grafting reaction of UPy-NCO. Due to the mono-coordination reaction between the pyridine ring and Cu^{2+} ions, new mesopores (8–20 nm) were introduced, enriching the pore architecture of the MOF@$Ti_3C_2T_X$ hybrid. Profiting from the enhanced grafting reaction between the $-NH_2$ groups and UPy-NCO units, quadruple hydrogen bonds among frameworks were increased, leading to the formation of hierarchical cross-linked networks. The as-obtained Cu-MOF@$Ti_3C_2T_X$-20%DAP-UPy presented a capacitance of 158 F g^{-1} at 0.2 A g^{-1}, delivering a capacitance retention of 88% at 5 A g^{-1}, while the Cu-MOF@$Ti_3C_2T_X$-UPy exhibited a capacitance of 128 F g^{-1} at 0.2 A g^{-1}, presenting a 73.6% retention at 5 A g^{-1}. Furthermore, the Cu-MOF@$Ti_3C_2T_X$-20%DAP-UPy delivered a 91.1% capacitance retention after 5000 cycles, which is much superior to that of Cu-MOF@$Ti_3C_2T_X$-UPy (79.9%), presenting an apparent increase in cycling performance.

The prepared ASC device presented a specific capacitance of 50 F g^{-1} at 0.2 A g^{-1}, an energy density of 17.9 Wh kg^{-1}, and a capacitance retention of 85% after 5000 cycles at 2 A g^{-1}. Our strategy not only provides ideas for designing novel and efficient MOFs, but also provides innovative ideas for exploring novel porous frameworks with self-healing ability.

5. Conclusions

In this study, the structural stability and electrochemical performance of two-dimensional MOF electrodes were optimized by introducing a quadruple hydrogen bonding strategy; 2,6-diaminopyridine (DAP) was employed as a moderating ligand involved in MOF growth, and the quadruple hydrogen bonding network was strengthened by the grafting reaction of its amino group with UPy-NCO. It was found that the mono-coordination of the pyridine ring with Cu^{2+} formed 8–20 nm new mesopores, which enriched the pore structure of MOF@$Ti_3C_2T_X$ composites; the modified material, Cu-MOF@$Ti_3C_2T_X$-20% DAP-UPy, reached a specific capacity of 15.5 mm at 0.2 A g^{-1} with a specific capacity of 158 F g^{-1} (23% enhancement), 88% capacity retention at 5 A g^{-1} (control 73.6%), and 91.1% capacity retention after 5000 cycles (control 79.9%); and the assembled asymmetric supercapacitor (ASC) had an energy density of 17.9 Wh kg^{-1} and 85% capacity retention after 5000 cycles. This strategy not only improves the conductivity and cycling stability of MOF electrodes by constructing a hierarchical cross-linking network, but also provides an innovative idea for the design of porous framework materials with a self-healing function.

Supplementary Materials: The following supporting information can be downloaded at: https://www.mdpi.com/article/10.3390/nano15110864/s1, Figure S1. SEM images of the samples. Figure S2. XRD patterns of the samples. Figure S3. (a) CV curves of AC and Cu-MOF@$Ti_3C_2T_X$-20%DAP-UPy at 50 mV s^{-1} and (b) GCD curve of AC at 1 A g^{-1}.

Author Contributions: Methodology, F.L. and P.Y.; Validation, B.Y., Q.C., M.Y., Y.N. and M.Z.; Writing—original draft, X.Q.; Writing—review & editing, S.L. All authors have read and agreed to the published version of the manuscript.

Funding: This work is supported by the National Natural Science Foundation of China (52073126, 52173020).

Data Availability Statement: The data presented in the study are available on request from the corresponding authors.

Conflicts of Interest: The authors declare no conflicts of interest.

References

1. Du, M.; Li, Q.; Zhao, Y.; Liu, C.-S.; Pang, H. A review of electrochemical energy storage behaviors based on pristine metal–organic frameworks and their composites. *Coord. Chem. Rev.* **2020**, *416*, 41. [CrossRef]
2. Khalil, I.E.; Fonseca, J.; Reithofer, M.R.; Eder, T.; Chin, J.M. Tackling orientation of metal-organic frameworks (MOFs): The quest to enhance MOF performance. *Coord. Chem. Rev.* **2023**, *481*, 215043. [CrossRef]
3. Wang, W.; Chen, D.; Li, F.; Xiao, X.; Xu, Q. Metal-organic-framework-based materials as platforms for energy applications. *Chem* **2024**, *10*, 86–133. [CrossRef]
4. Kang, L.; Liang, Q.; Abdul, Q.; Rashid, A.; Ren, X.; Ma, H. Preparation technology and preservation mechanism of γ-CD-MOFs biaological packaging film loaded with curcumin. *Food Chem.* **2023**, *420*, 136142. [CrossRef]
5. Gan, Z.; Hu, X.; Xu, X.; Zhang, W.; Zou, X.; Shi, J.; Zheng, K.; Arslan, M. A portable test strip based on fluorescent europium-based metal–organic framework for rapid and visual detection of tetracycline in food samples. *Food Chem.* **2021**, *354*, 129501. [CrossRef]
6. Wang, X.; Pan, Y.; Wang, X.; Guo, Y.; Ni, C.; Wu, J.; Hao, C. High performance hybrid supercapacitors assembled with multi-cavity nickel cobalt sulfide hollow microspheres as cathode and porous typha-derived carbon as anode. *Ind. Crops Prod.* **2022**, *189*, 115863. [CrossRef]
7. Marimuthu, M.; Arumugam, S.S.; Sabarinathan, D.; Li, H.; Chen, Q. Metal organic framework based fluorescence sensor for detection of antibiotics. *Trends Food Sci. Technol.* **2021**, *116*, 1002–1028. [CrossRef]

8. Liang, N.; Hu, X.; Zhang, X.; Li, W.; Guo, Z.; Huang, X.; Li, Z.; Zhang, R.; Shen, T.; Zou, X.; et al. Ratiometric Sensing for Ultratrac Tetracycline Using Electrochemically Active Metal–Organic Frameworks as Response Signals. *J. Agric. Food Chem.* **2023**, *7* 7584–7592. [CrossRef]
9. Chen, X.; Xu, J.; Li, Y.; Zhang, L.; Bi, N.; Gou, J.; Zhu, T.; Jia, L. A novel intelligently integrated MOF-based ratio fluorescenc sensor for ultra-sensitive monitoring of TC in water and food samples. *Food Chem.* **2023**, *405*, 134899. [CrossRef]
10. Wang, K.-B.; Xun, Q.; Zhang, Q. Recent progress in metal-organic frameworks as active materials for supercapacitors. *EnergyChem* **2020**, *2*, 100025. [CrossRef]
11. Shao, G.; Yu, R.; Chen, N.; Ye, M.; Liu, X.Y. Stretchable Supercapacitors: From Materials and Structures to Devices. *Small Method* **2021**, *5*, 2000853. [CrossRef] [PubMed]
12. Linares-Moreau, M.; Brandner, L.A.; Velásquez-Hernández, M.d.J.; Fonseca, J.; Benseghir, Y.; Chin, J.M.; Maspoch, D.; Doonan, C Falcaro, P. Fabrication of Oriented Polycrystalline MOF Superstructures. *Adv. Mater.* **2024**, *36*, 2309645. [CrossRef] [PubMed]
13. Wu, C.; Geng, P.; Zhang, G.; Li, X.; Pang, H. Synthesis of Conductive MOFs and Their Electrochemical Application. *Small* **2024**, *2* 2308264. [CrossRef]
14. Pramanik, B.; Sahoo, R.; Das, M.C. pH-stable MOFs: Design principles and applications. *Coord. Chem. Rev.* **2023**, *493*, 215301 [CrossRef]
15. Zhang, B.; Zhu, Z.; Wang, X.; Liu, X.; Kapteijn, F. Water Adsorption in MOFs: Structures and Applications. *Adv. Funct. Mater* **2024**, *34*, 2304788. [CrossRef]
16. Wei, D.; Zhang, L.; Wang, Y.; Qiu, S.; Luo, Y.; Zou, Y.; Xu, F.; Sun, L.; Chu, H. Recent progress on construction and applications o metal-organic frameworks-based materials for lithium-ion batteries and supercapacitors. *Carbon Neutralization* **2024**, *3*, 396–414 [CrossRef]
17. Liu, X.; Zhang, Y.; Guo, X.; Pang, H. Electrospun metal–organic framework nanofiber membranes for energy storage anc environmental protection. *Adv. Fiber Mater.* **2022**, *4*, 1463–1485. [CrossRef]
18. Sun, R.; Dou, M.; Chen, Z.; Wang, R.; Zheng, X.; Zhang, Y.; Zhou, C.; Menezes, P.W. Engineering strategies of metal-organi frameworks toward advanced batteries. *Battery Energy* **2023**, *2*, 20220064. [CrossRef]
19. Wu, X.-M.; Liu, M.-M.; Guo, H.-X.; Ying, S.-M.; Chen, Z.-X. Polyoxovanadate-based MOFs microsphere constructed from 3-D discrete nano-sheets as supercapacitor. *Chin. J. Struct. Chem.* **2021**, *40*, 994–998.
20. Kumar, Y.A.; Vignesh, S.; Ramachandran, T.; Kumar, K.D.; Al-Sehemi, A.G.; Moniruzzaman, M.; Oh, T.H. Solidifying the future Metal-organic frameworks in zinc battery development. *J. Energy Storage* **2024**, *97*, 112826. [CrossRef]
21. Huang, M.; Liang, Z.; Huang, J.; Wen, Y.; Zhu, Q.-L.; Wu, X. Introduction of Multicomponent Dyes into 2D MOFs: A Strategy to Fabricate White Light-Emitting MOF Composite Nanosheets. *ACS Appl. Mater. Interfaces* **2023**, *15*, 11131–11140. [CrossRef] [PubMed]
22. Iqbal, M.Z.; Shaheen, M.; Khan, M.W.; Siddique, S.; Farid, S.; Aftab, S.; Wabaidur, S.M. The rise of 2D conductive metal-organic framework: $Cu_3(HHTP)_2$ d-π MOF for integrated battery-supercapacitor hybrids. *Mater. Today Sustain.* **2023**, *22*, 100331 [CrossRef]
23. Li, S.; Zhang, L.; Ye, P.; Zhu, M.; Nie, Y.; Dai, Y.; Yang, F. Construction of Battery-Like Hierarchical MOF@MXene Heterostructures for Hybrid Supercapacitors. *Cryst. Growth Des.* **2024**, *24*, 7445–7454. [CrossRef]
24. Liu, Z.; Chen, J.; Que, M.; Zheng, H.; Yang, L.; Yuan, H.; Ma, Y.; Li, Y.; Yang, X. 2D Ti_3C_2T MXene/MOFs composites derived CoNi bimetallic nanoparticles for enhanced microwave absorption. *Chem. Eng. J.* **2022**, *450*, 138442. [CrossRef]
25. Hussain, N.; Abbas, Z.; Ansari, S.N.; Kedarnath, G.; Mobin, S.M. Phosphorization Engineering on a MOF-Derived Metal Phosphide Heterostructure (Cu/Cu_3P@NC) as an Electrode for Enhanced Supercapacitor Performance. *Inorg. Chem.* **2023**, *62*, 17083–17092. [CrossRef]
26. Wang, X.; Xu, Y.; Li, Y.; Li, Y.; Li, Z.; Zhang, W.; Zou, X.; Shi, J.; Huang, X.; Liu, C.; et al. Rapid detection of cadmium ions in meat by a multi-walled carbon nanotubes enhanced metal-organic framework modified electrochemical sensor. *Food Chem.* **2021**, *357*, 129762. [CrossRef]
27. Liu, S.; Zhang, M.; Chen, Q.; Ouyang, Q. Multifunctional Metal–Organic Frameworks Driven Three-Dimensional Folded Paper-Based Microfluidic Analysis Device for Chlorpyrifos Detection. *J. Agric. Food Chem.* **2024**, *72*, 14375–14385. [CrossRef]
28. Guo, T.; Zhou, D.; Liu, W.; Su, J. Recent advances in all-in-one flexible supercapacitors. *Sci. China Mater.* **2020**, *64*, 27–45. [CrossRef]
29. Li, S.; Cheng, Q.; Ye, P.; Zhang, Y.; Zhang, L.; Liu, F.; Qiu, H.; Qu, X.; Nie, Y. Hierarchical two-dimensional Ti-MOF derived from MXene for hybrid supercapacitor electrodes. *Appl. Organomet. Chem.* **2024**, *38*, e7547. [CrossRef]
30. Ghosh, A.; Fathima Thanutty Kallungal, S.; Ramaprabhu, S. 2D Metal-Organic Frameworks: Properties, Synthesis, and Applications in Electrochemical and Optical Biosensors. *Biosensors* **2023**, *13*, 123. [CrossRef]
31. Zheng, Y.; Sun, F.-Z.; Han, X.; Xu, J.; Bu, X.-H. Recent Progress in 2D Metal-Organic Frameworks for Optical Applications. *Adv. Opt. Mater.* **2020**, *8*, 2000110. [CrossRef]

32. Zhang, H.; Mei, H.; Qin, D.; Li, Z.; Hou, Z.; Lu, X.; Xu, B.; Sun, D. Conversion of Amorphous MOF Microspheres into a Nickel Phosphate Battery-Type Electrode Using the "Anticollapse" Two-Step Strategy. *Inorg. Chem.* **2021**, *60*, 17094–17102. [CrossRef] [PubMed]
33. Shi, B.; Zhang, X.; Li, W.; Liang, N.; Hu, X.; Xiao, J.; Wang, D.; Zou, X.; Shi, J. An intrinsic dual-emitting fluorescence sensing toward tetracycline with self-calibration model based on luminescent lanthanide-functionalized metal-organic frameworks. *Food Chem.* **2023**, *400*, 133995. [CrossRef]
34. Odarczenko, M.; Thakare, D.; Li, W.; Venkateswaran, S.P.; Sottos, N.R.; White, S.R. Sunlight-Activated Self-Healing Polymer Coatings. *Adv. Eng. Mater.* **2020**, *22*, 1901223. [CrossRef]
35. Platonova, E.O.; Vlasov, E.; Pavlov, A.A.; Kireynov, A.; Nelyub, V.A.; Polezhaev, A.V. Self-healing polyurethane based on a difuranic monomer from biorenewable source. *J. Appl. Polym. Sci.* **2019**, *136*, 47869. [CrossRef]
36. Chang, K.; Jia, H.; Gu, S.-Y. A transparent, highly stretchable, self-healing polyurethane based on disulfide bonds. *Eur. Polym. J.* **2019**, *112*, 822–831. [CrossRef]
37. Guo, M.; Pitet, L.M.; Wyss, H.M.; Vos, M.; Dankers, P.Y.W.; Meijer, E.W. Tough Stimuli-Responsive Supramolecular Hydrogels with Hydrogen-Bonding Network Junctions. *J. Am. Chem. Soc.* **2014**, *136*, 6969–6977. [CrossRef]
38. Li, G.; Li, S.; Ahmed, J.; Tian, W.; Li, L. Flexible perovskite photodetector with room-temperature self-healing capability without external trigger. *InfoMat* **2024**, *6*, e12594. [CrossRef]
39. Liu, H.; Huang, H.; Wu, X.; Peng, H.; Li, Z.; Hu, J.; Yu, Q. Effects of external multi-ions and wet-dry cycles in a marine environment on autogenous self-healing of cracks in cement paste. *Cem. Concr. Res.* **2019**, *120*, 198–206. [CrossRef]
40. Guo, M.; Li, W.; Han, N.; Wang, J.; Su, J.; Li, J.; Zhang, X. Novel Dual-Component Microencapsulated Hydrophobic Amine and Microencapsulated Isocyanate Used for Self-Healing Anti-Corrosion Coating. *Polymers* **2018**, *10*, 319. [CrossRef]
41. Li, H.; Xin, L.; Gao, J.; Shao, Y.; Zhang, Z.; Ren, L. Underwater Bionic Self-Healing Superhydrophobic Coating with the Synergetic Effect Of Hydrogen Bonds and Self-Formed Bubbles. *Small* **2024**, *20*, 2309012. [CrossRef]
42. Das, M.; Baran Bhattacharya, A.; Rahman Parathodika, A.; Naskar, K. Room temperature Self-healable and extremely stretchable elastomer with improved mechanical Properties: Exploring a simplistic Metal-Ligand interaction. *Eur. Polym. J.* **2022**, *174*, 111341. [CrossRef]
43. Rahimpour, S.; Luo, L.; Teimuri-Mofrad, R. Preparation of ferrocenyl-furan modified graphene oxide via Diels-Alder click reaction and using of its polypyrrole nanocomposites as supercapacitor electrode material. *Electrochim. Acta* **2022**, *416*, 140285. [CrossRef]
44. Sun, Z.; Zhang, J.; Ye, F.; Wang, W.; Wang, G.; Zhang, Z.; Li, S.; Zhou, Y.; Cai, J. Vulcanization treatment: An effective way to improve the electrochemical cycle stability of polyaniline in supercapacitors. *J. Power Sources* **2019**, *443*, 227246. [CrossRef]
45. Wang, P.; Wang, Z.; Zhang, X.; Liao, Y.; Duan, W.; Yue, Y.; Zhang, Y. Stretchable Superhydrophobic Supercapacitor with Excellent Self-Healing Ability. *Energy Fuels* **2023**, *37*, 5567–5576. [CrossRef]
46. Tee, B.C.K.; Wang, C.; Allen, R.; Bao, Z. An electrically and mechanically self-healing composite with pressure- and flexion-sensitive properties for electronic skin applications. *Nat. Nanotechnol.* **2012**, *7*, 825–832. [CrossRef]
47. Nakahata, M.; Takashima, Y.; Yamaguchi, H.; Harada, A. Redox-responsive self-healing materials formed from host–guest polymers. *Nat. Commun.* **2011**, *2*, 511. [CrossRef] [PubMed]
48. Hang, X.; Xue, Y.; Cheng, Y.; Du, M.; Du, L.; Pang, H. From Co-MOF to CoNi-MOF to Ni-MOF: A Facile Synthesis of 1D Micro-/Nanomaterials. *Inorg. Chem.* **2021**, *60*, 13168–13176. [CrossRef]
49. Mashkoor, F.; Lee, S.J.; Yi, H.; Noh, S.M.; Jeong, C. Self-Healing Materials for Electronics Applications. *Int. J. Mol. Sci.* **2022**, *23*, 622. [CrossRef]
50. Mathis, T.S.; Kurra, N.; Wang, X.; Pinto, D.; Simon, P.; Gogotsi, Y. Energy Storage Data Reporting in Perspective—Guidelines for Interpreting the Performance of Electrochemical Energy Storage Systems. *Adv. Energy Mater.* **2019**, *9*, 1902007. [CrossRef]
51. Sun, J.; Liu, Y.; Wu, G.; Zhang, Y.; Zhang, R.; Li, X.J. A Fusion Parameter Method for Classifying Freshness of Fish Based on Electrochemical Impedance Spectroscopy. *J. Food Qual.* **2021**, *2021*, 6664291. [CrossRef]
52. Li, Y.; Meng, S.; Dong, N.; Wei, Y.; Wang, Y.; Li, X.; Liu, D.; You, T. Space-Confined Electrochemical Aptasensing with Conductive Hydrogels for Enhanced Applicability to Aflatoxin B1 Detection. *J. Agric. Food Chem.* **2023**, *71*, 14806–14813. [CrossRef] [PubMed]
53. Liu, S.; Meng, S.; Wang, M.; Li, W.; Dong, N.; Liu, D.; Li, Y.; You, T. In-depth interpretation of aptamer-based sensing on electrode: Dual-mode electrochemical-photoelectrochemical sensor for the ratiometric detection of patulin. *Food Chem.* **2023**, *410*, 135450. [CrossRef] [PubMed]
54. Poonam; Sharma, K.; Arora, A.; Tripathi, S.K. Review of supercapacitors: Materials and devices. *J. Energy Storage* **2019**, *21*, 801–825. [CrossRef]
55. Bi, S.; Banda, H.; Chen, M.; Niu, L.; Chen, M.; Wu, T.; Wang, J.; Wang, R.; Feng, J.; Chen, T.; et al. Molecular understanding of charge storage and charging dynamics in supercapacitors with MOF electrodes and ionic liquid electrolytes. *Nat. Mater.* **2020**, *19*, 552–558. [CrossRef]

56. Acharya, D.; Pathak, I.; Dahal, B.; Lohani, P.C.; Bhattarai, R.M.; Muthurasu, A.; Kim, T.; Ko, T.H.; Chhetri, K.; Kim, H.Y. Immoderate nanoarchitectures of bimetallic MOF derived Ni–Fe–O/NPC on porous carbon nanofibers as freestanding electrode for asymmetric supercapacitors. *Carbon* **2023**, *201*, 12–23. [CrossRef]
57. Kim, M.; Xin, R.; Earnshaw, J.; Tang, J.; Hill, J.P.; Ashok, A.; Nanjundan, A.K.; Kim, J.; Young, C.; Sugahara, Y.; et al. MOF-derived nanoporous carbons with diverse tunable nanoarchitectures. *Nat. Protoc.* **2022**, *17*, 2990–3027. [CrossRef]
58. Wang, C.; Kim, J.; Tang, J.; Kim, M.; Lim, H.; Malgras, V.; You, J.; Xu, Q.; Li, J.; Yamauchi, Y. New strategies for novel MOF-derived carbon materials based on nanoarchitectures. *Chem* **2020**, *6*, 19–40. [CrossRef]
59. Zheng, S.; Sun, Y.; Xue, H.; Braunstein, P.; Huang, W.; Pang, H. Dual-ligand and hard-soft-acid-base strategies to optimize metal-organic framework nanocrystals for stable electrochemical cycling performance. *Natl. Sci. Rev.* **2022**, *9*, nwab197. [CrossRef]
60. Zhao, L.; Yang, J.; Gong, M.; Li, K.; Gu, J. Specific Screening of Prostate Cancer Individuals Using an Enzyme-Assisted Substrate Sensing Platform Based on Hierarchical MOFs with Tunable Mesopore Size. *J. Am. Chem. Soc.* **2021**, *143*, 15145–15151. [CrossRef]
61. Wang, S.; Wang, J.; Zeng, M.; Yang, J.; Hu, N.; Su, Y.; Zhou, Z.; Pang, H.; Yang, Z. Synthesis of nickel-metal organic framework nanoplates with pyridine modulation and application to supercapacitors. *J. Energy Storage* **2021**, *38*, 102528. [CrossRef]
62. Li, Z.-X.; Yang, B.-L.; Zou, K.-Y.; Kong, L.; Yue, M.-L.; Duan, H.-H. Novel porous carbon nanosheet derived from a 2D Cu-MOF: Ultrahigh porosity and excellent performances in the supercapacitor cell. *Carbon* **2019**, *144*, 540–548. [CrossRef]
63. Otun, K.O.; Zong, S.; Hildebrandt, D.; Liu, X. Self-assembled Zn-functionalized Ni-MOF as an efficient electrode for electrochemical energy storage. *J. Phys. Chem. Solids* **2022**, *167*, 110779. [CrossRef]
64. Jo, Y.H.; Zhou, B.; Jiang, K.; Li, S.; Zuo, C.; Gan, H.; He, D.; Zhou, X.; Xue, Z. Self-healing and shape-memory solid polymer electrolytes with high mechanical strength facilitated by a poly (vinyl alcohol) matrix. *Polym. Chem.* **2019**, *10*, 6561–6569. [CrossRef]
65. Saraf, M.; Rajak, R.; Mobin, S.M. A fascinating multitasking Cu-MOF/rGO hybrid for high performance supercapacitors and highly sensitive and selective electrochemical nitrite sensors. *J. Mater. Chem. A* **2016**, *4*, 16432–16445. [CrossRef]
66. Hussain, I.; Kathiresan, M.; Singh, K.; Kalidasan, B.; Mendhe, A.C.; Islam, M.N.; Meng, K.; Aslam, M.K.; Hanif, M.B.; Al Zoubi, W.; et al. Interface and surface engineering of MXenes and COFs for energy storage and conversion. *InfoMat* **2025**, e70011. [CrossRef]
67. Wang, J.; Du, C.F.; Xue, Y.; Tan, X.; Kang, J.; Gao, Y.; Yu, H.; Yan, Q. MXenes as a versatile platform for reactive surface modification and superior sodium-ion storages. *Exploration* **2021**, *1*, 20210024. [CrossRef]
68. Nguyen, D.K.; Schepisi, I.M.; Amir, F.Z. Extraordinary cycling stability of $Ni_3(HITP)_2$ supercapacitors fabricated by electrophoretic deposition: Cycling at 100,000 cycles. *Chem. Eng. J.* **2019**, *378*, 122150. [CrossRef]
69. Li, W.-H.; Ding, K.; Tian, H.-R.; Yao, M.-S.; Nath, B.; Deng, W.-H.; Wang, Y.; Xu, G. Conductive Metal–Organic Framework Nanowire Array Electrodes for High-Performance Solid-State Supercapacitors. *Adv. Funct. Mater.* **2017**, *27*, 1702067. [CrossRef]
70. Zhang, X.; Qu, N.; Yang, S.; Lei, D.; Liu, A.; Zhou, Q. Cobalt induced growth of hollow MOF spheres for high performance supercapacitors. *Mater. Chem. Front.* **2021**, *5*, 482–491. [CrossRef]
71. Wang, Y.; Liu, Y.; Wang, H.; Liu, W.; Li, Y.; Zhang, J.; Hou, H.; Yang, J. Ultrathin NiCo-MOF Nanosheets for High-Performance Supercapacitor Electrodes. *ACS Appl. Energy Mater.* **2019**, *2*, 2063–2071. [CrossRef]
72. Wang, Y.; Nie, S.; Liu, Y.; Yan, W.; Lin, S.; Cheng, G.; Yang, H.; Luo, J. Room-Temperature Fabrication of a Nickel-Functionalized Copper Metal–Organic Framework (Ni@Cu-MOF) as a New Pseudocapacitive Material for Asymmetric Supercapacitors. *Polymers* **2019**, *11*, 821. [CrossRef] [PubMed]
73. Singh, M.K.; Gupta, A.K.; Krishnan, S.; Guha, N.; Marimuthu, S.; Rai, D.K. A new hierarchically porous Cu-MOF composited with rGO as an efficient hybrid supercapacitor electrode material. *J. Energy Storage* **2021**, *43*, 103301. [CrossRef]
74. Jayaramulu, K.; Horn, M.; Schneemann, A.; Saini, H.; Bakandritsos, A.; Ranc, V.; Petr, M.; Stavila, V.; Narayana, C.; Scheibe, B.; et al. Covalent Graphene-MOF Hybrids for High-Performance Asymmetric Supercapacitors. *Adv. Mater.* **2021**, *33*, 2004560. [CrossRef] [PubMed]
75. Zhong, Y.; Cao, X.; Ying, L.; Cui, L.; Barrow, C.; Yang, W.; Liu, J. Homogeneous nickel metal-organic framework microspheres on reduced graphene oxide as novel electrode material for supercapacitors with outstanding performance. *J. Colloid Interface Sci.* **2020**, *561*, 265–274. [CrossRef]
76. Ren, Y.-F.; He, Z.-L.; Zhao, H.-Z.; Zhu, T. Fabrication of MOF-derived mixed metal oxides with carbon residues for pseudocapacitors with long cycle life. *Rare Met.* **2022**, *41*, 830–835. [CrossRef]
77. Han, Y.; Hou, X.-Y.; Wang, X.; Fu, F.; Tang, L.; Wang, J.-J. A Capacity Supercapacitor Electrode Material of Ni-MOF with High Surface Area and Porosity. *Chin. J. Struct. Chem.* **2019**, *38*, 1779–1786.
78. Kim, M.; Nara, H.; Asakura, Y.; Hamada, T.; Yan, P.; Earnshaw, J.; An, M.; Eguchi, M.; Yamauchi, Y. End-to-End Pierced Carbon Nanosheets with Meso-Holes. *Adv. Sci.* **2025**, *12*, 2409546. [CrossRef]

nanomaterials MDPI

Article

Engineering Nonvolatile Polarization in 2D α-In_2Se_3/α-Ga_2Se_3 Ferroelectric Junctions

Peipei Li, Delin Kong, Jin Yang, Shuyu Cui, Qi Chen, Yue Liu, Ziheng He, Feng Liu, Yingying Xu, Huiyun Wei, Xinhe Zheng * and Mingzeng Peng *

Beijing Key Laboratory for Magneto-Photoelectrical Composite and Interface Science, School of Mathematics and Physics, University of Science and Technology Beijing, No. 30, Xueyuan Road, Beijing 100083, China; m202210750@xs.ustb.edu.cn (P.L.); d202410764@xs.ustb.edu.cn (D.K.); d202210420@xs.ustb.edu.cn (J.Y.); m202310753@xs.ustb.edu.cn (S.C.); u202242156@xs.ustb.edu.cn (Q.C.); m202310770@xs.ustb.edu.cn (Y.L.); u202242160@xs.ustb.edu.cn (Z.H.); u202242159@xs.ustb.edu.cn (F.L.); xuyingying@ustb.edu.cn (Y.X.); huiyunwei@ustb.edu.cn (H.W.)

* Correspondence: xinhezheng@ustb.edu.cn (X.Z.); mzpeng@ustb.edu.cn (M.P.)

Abstract: The advent of two-dimensional (2D) ferroelectrics offers a new paradigm for device miniaturization and multifunctionality. Recently, 2D α-In_2Se_3 and related III–VI compound ferroelectrics manifest room-temperature ferroelectricity and exhibit reversible spontaneous polarization even at the monolayer limit. Here, we employ first-principles calculations to investigate group-III selenide van der Waals (vdW) heterojunctions built up by 2D α-In_2Se_3 and α-Ga_2Se_3 ferroelectric (FE) semiconductors, including structural stability, electrostatic potential, interfacial charge transfer, and electronic band structures. When the FE polarization directions of α-In_2Se_3 and α-Ga_2Se_3 are parallel, both the α-In_2Se_3/α-Ga_2Se_3 P↑↑ (UU) and α-In_2Se_3/α-Ga_2Se_3 P↓↓ (NN) configurations possess strong built-in electric fields and hence induce electron–hole separation, resulting in carrier depletion at the α-In_2Se_3/α-Ga_2Se_3 heterointerfaces. Conversely, when they are antiparallel, the α-In_2Se_3/α-Ga_2Se_3 P↓↑ (NU) and α-In_2Se_3/α-Ga_2Se_3 P↑↓ (UN) configurations demonstrate the switchable electron and hole accumulation at the 2D ferroelectric interfaces, respectively. The nonvolatile characteristic of ferroelectric polarization presents an innovative approach to achieving tunable n-type and p-type conductive channels for ferroelectric field-effect transistors (FeFETs). In addition, in-plane biaxial strain modulation has successfully modulated the band alignments of the α-In_2Se_3/α-Ga_2Se_3 ferroelectric heterostructures, inducing a type III–II–III transition in UU and NN, and a type I–II–I transition in UN and NU, respectively. Our findings highlight the great potential of 2D group-III selenides and ferroelectric vdW heterostructures to harness nonvolatile spontaneous polarization for next-generation electronics, nonvolatile optoelectronic memories, sensors, and neuromorphic computing.

Keywords: group-III selenides; 2D ferroelectrics; polarization engineering; band alignments; strain modulation

1. Introduction

With the advancement in semiconductor technology, high-density storage, and computing integrated applications, ferroelectric materials and devices are continuously pursuing miniaturization. However, as dimensions shrink, traditional ferroelectrics, including $BiFeO_3$ [1], $BaTiO_3$ [2], $PbTiO_3$ [3], etc., encounter the challenge of size effects. As a result, the exploration of ultrathin ferroelectric materials remains the key issue in this field.

Two-dimensional ferroelectrics possess inherently stable layered structures, characterized by strong intralayer bonding, weak interlayer vdW interactions, and high electron mobility due to their unique quantum size effects. These attributes offer novel opportunities for advancing the exploration of 2D ferroelectricity and its extensive applications at the atomic scale. The symmetry breaking of 2D ferroelectrics has been achieved through various strategies, such as stacking vdW heterostructures, applying strains, and engineering Janus structures [4,5]. Recently, 2D ferroelectrics such as 1*T*-MoS_2 [6], α-In_2Se_3 [7], $CuInP_2S_6$ [4], and MX (M = Ge, Sn; X = S, Se, Te) [8] have gained tremendous interest due to their unique features in contrast to 3D counterparts. The breakthrough of 2D ferroelectricity has driven emerging applications in 2D ferroelectric field-effect transistors [9,10], ferroelectric tunnel junctions [11], and high-photosensitivity and rapid-response photodetectors [12]. As we are entering into the post-Moore era, the discovery of 2D ferroelectrics with desired performance characteristics becomes crucial for advancing the frontiers of electronic device technology.

As part of the 2D ferroelectric family, 2D III–VI layered semiconductors have garnered extensive attention. Specifically, they exhibit five distinct structural phases (α, β, γ, δ, κ). Among these, α, β, and γ are common phases [13,14], and δ and κ are uncommon phases [15,16]. The γ phase is identified as a three-dimensional material with a hexagonal defect wurtzite crystal structure. In contrast, the other phases including α, β, δ, and κ exhibit a layered structure [17]. For instance, α-In_2Se_3 is the most stable layered structure with spontaneous ferroelectric polarization at room temperature [7]. Two-dimensional α-In_2Se_3 has been experimentally and theoretically confirmed to possess a robust out-of-plane ferroelectric and piezoelectricity polarization [18–20]. The monolayer α-In_2Se_3 is alternately composed of Se and In atomic layers through covalent bonding to form a quintuple-layer structure of Se–In–Se–In–Se [21]. The unequal interlayer distances between the middle Se layer and its adjacent In layers lead to the breaking of symmetry. Several research groups have subsequently confirmed the spontaneous in-plane and out-of-plane ferroelectricity of ultrathin α-In_2Se_3 at room temperature [19,22] and fabricated high-performance rigid and flexible optoelectronic detectors [23]. Similar to other polar materials, symmetry breaking in α-In_2Se_3 leads to the misalignment of positive and negative charge centers, producing a dipole moment and spontaneous polarization. The out-of-plane intrinsic polarization is reversible, permitting the manipulation of ferroelectric polarization in both upward and downward directions by laterally shifting the central Se layer. When integrated with other polar semiconductors, 2D α-In_2Se_3 can induce charge redistribution and further manipulate electronic energy properties. It was found that the interfacial dipole interaction plays a crucial role in the external electric field modulation, with a direct correlation to the charge distribution at α-In_2Se_3/GaN polar heterointerfaces [24]. The asymmetric metal/α-In_2Se_3/Si crossbar ferroelectric semiconductor junctions have been demonstrated to enhance the modulation of the effective Schottky barrier height by the ferroelectric polarization [25]. The inherent nonvolatile ferroelectric switching and semiconducting characteristics of α-In_2Se_3 show versatile fields including nonvolatile memory, neuromorphic computing, optoelectronics, and thermoelectric applications [15,26]. As another member of 2D III–VI layered semiconductors, α-Ga_2Se_3 has been demonstrated to possess excellent structural stability, which is obtained by replacing In of α-In_2Se_3 with Ga. The first-principles calculations have indicated that α-Ga_2Se_3 similar to α-In_2Se_3 is a semiconductor with Young's modulus of less than 100 N m^{-1} in a deformation range of up to about 30% [27]. The switched polarization was observed in 2D α-Ga_2Se_3 nanoflakes of approximately 4 nm with a high switching temperature of up to 450 K [28]. Such polarization switching could arise from the displacement of Ga vacancy between neighboring asymmetrical sites by applying an electric field [28]. In the graphene/α-Ga_2Se_3 vdW heterojunctions, the presence of spontaneous polarization reveals intriguing interfacial characteristics by controlling contact types and external disturbances for the multifunctional design of high-performance Schottky devices [29]. The ferroelectric nature of α-Ga_2Se_3 has opened avenues for its use in nonvolatile memory devices and non-linear optical applications. Other studies have also reported that α-Ga_2Se_3 exhibits significant piezoelectric performance, showing broad prospects in ultrathin piezoelectric sensors and nanogenerators [30]. Although the α-In_2Se_3 and α-Ga_2Se_3 layered materials have been extensively studied, the limitation of 2D ferroelectric multilayers and complex stacking heterostructures impedes a comprehensive understanding of their intricate polarization interactions, polarity modulation mechanisms, and synergistic effects. Therefore, the exploration of 2D group-III selenide ferroelectrics and polar vdW heterostructures holds profound significance, extending 2D ferroelectrics and semiconducting optoelectronics by offering novel functionalities and characteristics that surpass those of traditional 3D ferroelectrics.

In this study, we have proposed 2D α-In_2Se_3/α-Ga_2Se_3 ferroelectric vdW heterostructures and investigated the ferroelectric polarization modulation by first-principles calculations. Considering the reversible ferroelectric polarization of α-In_2Se_3 and α-Ga_2Se_3, the positive and negative c-axes are defined as polarity up (P↑) and polarity down (P↓), respectively. In terms of polarity manipulation, there are switchable ferroelectric polarization configurations including α-In_2Se_3/α-Ga_2Se_3 P↑↑ (UU), α-In_2Se_3/α-Ga_2Se_3 P↑↓ (UN), α-In_2Se_3/α-Ga_2Se_3 P↓↑ (NU), and α-In_2Se_3/α-Ga_2Se_3 P↓↓ (NN). Our calculations demonstrate that the UN and NU configurations exhibit a pronounced accumulation of holes and electrons at the ferroelectric heterointerfaces of α-In_2Se_3/α-Ga_2Se_3, facilitating ferroelectric modulation of conductivity types between the p-channel and n-channel within FeFETs. In addition, the electronic band structures of the α-In_2Se_3/α-Ga_2Se_3 polar heterostructures have been effectively manipulated by employing in-plane biaxial strain modulation. From compressive to tensile strains, their band alignments undergo a type III–II–III transition in UU and NN configurations, and a type I–II–I transition in UN and NU configurations. Strain engineering provides a versatile approach for controllable energy band transformation, facilitating the charge separation, transfer, recombination, and redistribution processes in 2D α-In_2Se_3/α-Ga_2Se_3 ferroelectric heterostructures. Therefore, 2D group-III selenides and ferroelectric vdW heterostructures provide atomic-level platforms that integrate electronic, optical, and memory functionalities for nonvolatile multifunctional applications.

2. Calculation Methods

First-principles calculations were carried out using the Vienna ab initio Simulation Package (VASP) simulation method based on density functional theory (DFT) [31,32]. To calculate the electron exchange–correlation interactions, we employed the meta-generalized gradient approximation (meta-GGA) and the Strongly Constrained and Appropriately Normed (SCAN) functional to calculate the geometry optimizations and electronic band structures [33,34]. Compared to the local density approximation (LDA) and the Perdew, Burke, and Ernzerhof (PBE) method, meta-GGA SCAN has been tested in diversely bonded systems, where it was shown to be sophisticated enough to model a wide range of physical structures without being fitted to any bonded system [35]. It can describe an intermediate range of dispersion via the kinetic energy density and can be proven to deliver sufficiently accurate ground-state properties for diversely bonded systems [36,37]. The projector-augmented wave (PAW) method was used to describe the ion–electron interactions [38,39]. For the calculations, we constructed α-In_2Se_3/α-Ga_2Se_3 (hexagonal, space group $R3m$) unit cells with different polarity configurations, as depicted in Figure 1. Their 2D ferroelectric polarization modulation has been investigated on the structural stability, electrostatic potential, interfacial charge transfer, and electronic band structures by both the ferroelectric polarization reversals of α-In_2Se_3 and α-Ga_2Se_3, which serve as a foundation for the exploration of 2D ferroelectric/optoelectronic multifunctional applications. We utilized a plane-wave basis expansion with a cutoff energy of E_{cut} = 500 eV and employed a 12 × 12 × 1 Monkhorst–Pack K point grid in the Brillouin Zone. And, Grimme's DFT-D3 vdW corrections were incorporated based on the semiempirical GGA-type theory [40]. A vacuum layer of 15 Å along the c direction was adopted to ensure decoupling between periodically repeated systems. For the relaxation of the unit cells at zero strain, all atoms were fully relaxed with a convergence accuracy of 1×10^{-8} eV. The interaction force between atoms was found to be less than |0.02| eV/Å. When considering the strain modulation on the α-In_2Se_3/α-Ga_2Se_3 ferroelectric heterojunctions, the in-plane biaxial strains (ε) in the a and b directions are defined as follows: $\varepsilon = 100\% \times (a_1 - a_0)/a_0$, where a_1 and a_0 represent the lattice parameters with and without applied biaxial strains, respectively.

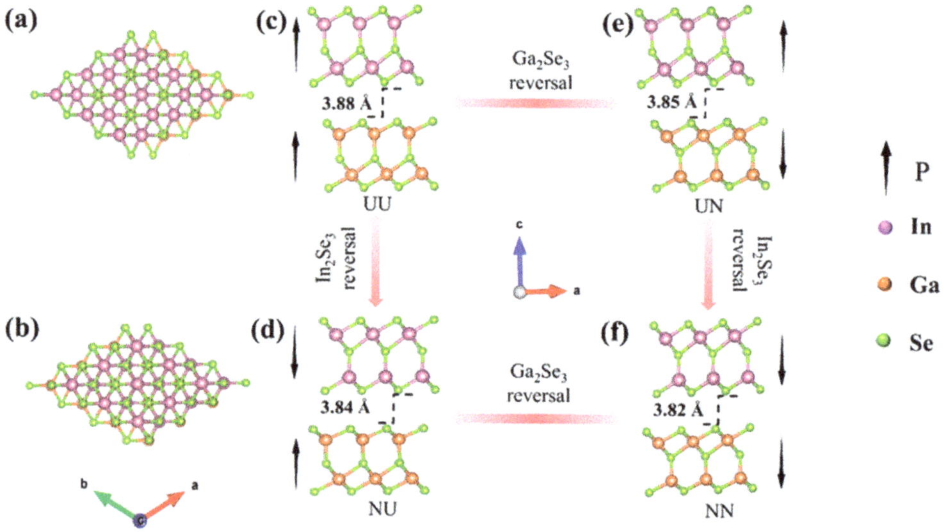

Figure 1. Two-dimensional α-In_2Se_3/α-Ga_2Se_3 ferroelectric heterostructures: top views of (**a**) UU and UN, (**b**) NU and NN, and side views of (**c**) UU, (**d**) NU, (**e**) UN, and (**f**) NN. The purple, orange, and green represent In, Ga, and Se atoms, respectively. The black arrow represents the direction of polarization.

3. Results and Discussion

3.1. Polarity Configuration and Structural Stability

To investigate the electronic properties of 2D α-In_2Se_3/α-Ga_2Se_3 ferroelectric heterostructures, the fully-optimized lattice constants are a = b = 4.03 Å for α-In_2Se_3 and a = b = 3.84 Å for α-Ga_2Se_3, respectively, which are in good agreement with previous reports [9,41]. The small lattice mismatch between α-In_2Se_3 and α-Ga_2Se_3 is 4.8%, demonstrating good feasibility to construct their vdW heterointerfaces with low strains and few defects. In addition, different stacking orders between α-In_2Se_3/α-Ga_2Se_3 need to be considered, as shown in Figure 1, including the In–Se, In–Ga, and Se–Se alignments. The structural stability of each stacking order has been examined by calculating the binding energy (E_b) of 2D α-In_2Se_3/α-Ga_2Se_3 ferroelectric heterostructure systems as follows:

$$E_b = E_{In2Se3/Ga2Se3} - E_{In2Se3} - E_{Ga2Se3},$$

where $E_{In2Se3/Ga2Se3}$, E_{In2Se3}, and E_{Ga2Se3} are the total energies of α-In_2Se_3/α-Ga_2Se_3 heterostructures, α-In_2Se_3, and α-Ga_2Se_3, respectively. As shown in Figure 1a,b, it is found that the In–Ga alignment has the smaller binding energy in α-In_2Se_3 (P↑)/α-Ga_2Se_3 configurations, while the In–Se alignment has the smaller binding energy in α-In_2Se_3 (P↓)/α-Ga_2Se_3 configurations. The stacking order has a direct change from In–Ga to In–Se due to switching the α-In_2Se_3 polarization from upward to downward. Based upon the above results, the four most stable α-In_2Se_3/α-Ga_2Se_3 ferroelectric heterostructures have been achieved in Figure 1c–f by stacking α-In_2Se_3 (P↑) on α-Ga_2Se_3 (P↑) (UU), α-In_2Se_3 (P↑) on α-Ga_2Se_3 (P↓) (UN), α-In_2Se_3 (P↓) on α-Ga_2Se_3 (P↑) (NU), and α-In_2Se_3 (P↓) on α-Ga_2Se_3 (P↓) (NN), respectively. It should be noted that the NN configuration has the shortest vdW interlayer distance. Therefore, it demonstrates that the stacking order in combination with ferroelectric polarization plays an important influence in 2D interfacial interactions, such as polarization coupling, electrostatic interaction, charge transfer, photoelectric conversion, etc. [42,43].

3.2. Band Structures and Density of States

As shown in Figure 2a,b, the projected electronic band structures and density of states (DOS) of monolayer α-In_2Se_3 and α-Ga_2Se_3 have been firstly calculated by using the SCAN function. Both of them have similar band structures with the conduction band minimum (CBM) located at the Γ point and the valence band maximum (VBM) positioned between the Γ and M points. The bandgaps of α-In_2Se_3 and α-Ga_2Se_3 are 1.14 eV and 1.27 eV, respectively, and both exhibit an indirect bandgap. According to the DOS results, the electronic states at both the CBM and VBM are predominately attributed to the Se atoms, in good consistence with the previous results [23].

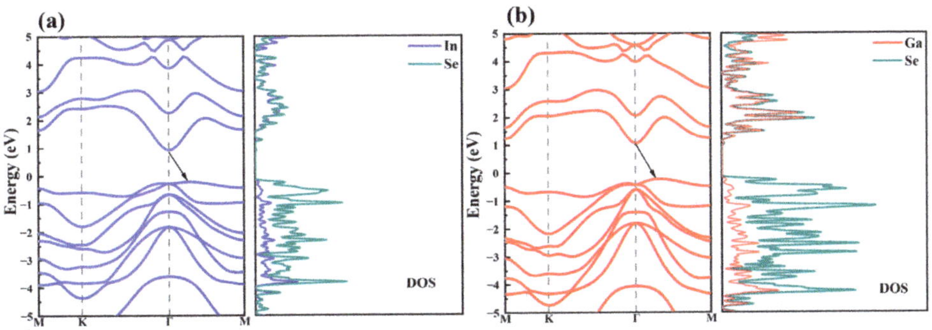

Figure 2. Projected band structures and DOS of (**a**) α-In_2Se_3 and (**b**) α-Ga_2Se_3. The Fermi level is set as 0.

When two different materials (namely, A and B) contact each other, the band alignment types are based on the classification of semiconductor heterostructures. They are categorized into three types: type I (Straddling gap), type II (staggered gap), and type III (broken gap). In type I, the conduction band minimum (CBM) of band B is higher in energy than the CBM of band A, and the valence band maximum (VBM) of band B is lower in energy than the VBM of band A. This results in a straddling gap where the energy levels of bands A and B overlap. In type II, both the CBM and VBM of band B are higher in energy than the CBM and VBM of band A, respectively. A staggered gap is formed where their energy levels do not overlap. In type III, the conduction band minimum of band A is lower in energy than the VBM of band B, resulting in a broken gap. To further clarify the ferroelectric polarization interactions, Figure 3 shows the projected band structures and DOS of 2D ferroelectric α-In_2Se_3/α-Ga_2Se_3 heterostructures. The red and blue lines correspond to the projected weights of electrons for α-Ga_2Se_3 and α-In_2Se_3, respectively.

By switching the ferroelectric polarization state of α-Ga_2Se_3 from P↑ to P↓, the electronic energy bands of α-Ga_2Se_3 and α-In_2Se_3 experience pronounced downward and upward shifts in opposite trends, respectively. They are both observed during the configuration transitions from UU to UN in Figure 3a,b and from NU to NN in Figure 3c,d. Consistently, upon reversing the ferroelectric polarization state of α-In_2Se_3 from P↑ to P↓, the electronic energy bands of α-Ga_2Se_3 and α-In_2Se_3 undergo similar bandshifts from UU to NU or from UN to NN, as depicted in Figure 3a and c or Figure 3b and d, respectively. The UU, UN, NU, and NN configurations have indirect bandgaps of 0.51 eV, 0.85 eV, 0.63 eV, and 0.08 eV, respectively. Although all of them exhibit type-II alignments, the CBM and VBM originate from different ferroelectric layers of α-In_2Se_3 and α-Ga_2Se_3. Specifically, the heterojunction bands come from the CBM of α-In_2Se_3 and the VBM of α-Ga_2Se_3 in UU and from the VBM of α-In_2Se_3 and the CBM of α-Ga_2Se_3 in NN. As a result, the tunable type-II alignments can not only drive the spontaneous separation of free electrons and holes but also determine the charge transfer directions by reversing the 2D ferroelectric polarization field. It is noted that the CBM of α-Ga_2Se_3 and VBM of α-In_2Se_3 in NN approach the Fermi level, resulting in the largest bandshifts, as shown in Figure 3d. This highlights the significant role of polarity manipulation in controlling the electronic energy bands of 2D ferroelectric heterojunctions.

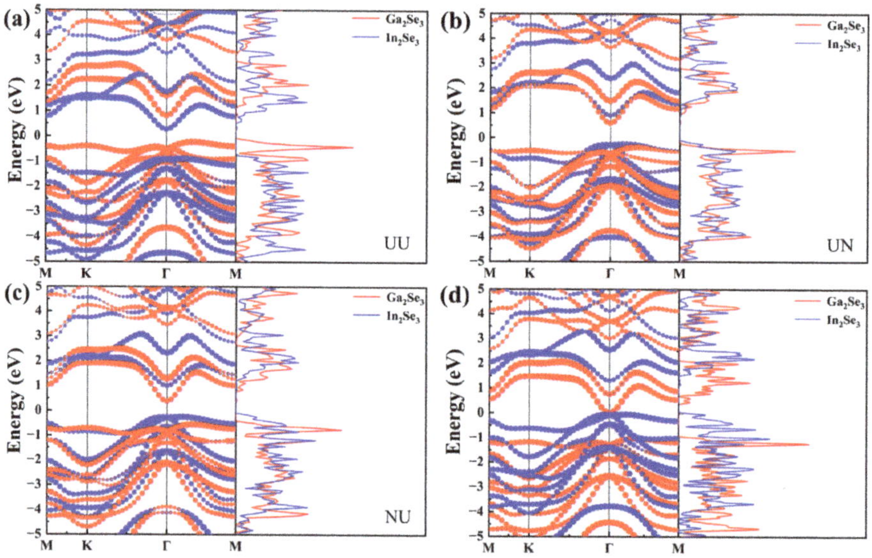

Figure 3. Projected band structures and DOS of the α-In_2Se_3/α-Ga_2Se_3 heterojunctions in (**a**) UU, (**b**) UN, (**c**) NU, and (**d**) NN. The blue and red marks represent the contributions of the α-In_2Se_3 and α-Ga_2Se_3 layers on projected band structures and projected density of states, respectively. The Fermi level is set as 0.

3.3. Electrostatic Potential and Charge Transfer

The electrostatic potential and charge transfer distribution are intricately linked to the ferroelectric polarization interactions at the 2D α-In_2Se_3/α-Ga_2Se_3 heterointerfaces, which in turn significantly impact their electronic energy bands and electrical transport behaviors. Figure 4 calculated the electrostatic potential distribution along the z-axis for α-In_2Se_3, α-Ga_2Se_3, and α-In_2Se_3/α-Ga_2Se_3 ferroelectrics in four stacking configurations, respectively. As an indicator of ferroelectric polarization strength, the electrostatic potential differences (ΔΦ) are 1.32 eV and 1.11 eV for α-In_2Se_3 and α-Ga_2Se_3, as shown in Figure 4a and b, respectively. So, α-In_2Se_3 has a more robust ferroelectric polarization than α-Ga_2Se_3, in consistency with the different electronegativity between In and Ga. After contacting each other, the α-In_2Se_3/α-Ga_2Se_3 heterostructures in Figure 4c–f obtain the electrostatic potential differences of 2.07 eV (UU), 0.22 eV (UN), 0.24 eV (NU), and 1.69 eV (NN), respectively. In the UU and UN configurations, the electrostatic potential of α-Ga_2Se_3 P↑ or P↓ is higher than α-In_2Se_3, indicating that the polarization-induced electric field is directed from α-In_2Se_3 to α-Ga_2Se_3, as shown in Figure 4c,d. As shown in Figure 4e,f, when switching the ferroelectric polarization state of the α-In_2Se_3 layer from P↑ to P↓ in the NU and NN configurations, the electrostatic potential of α-Ga_2Se_3 P↑ or P↓ becomes lower than α-In_2Se_3, thereby reversing the direction of the polarization-induced electric field. When α-In_2Se_3 and α-Ga_2Se_3 are polarized in the same directions, the electrostatic potential differences of the UU and NN configurations are higher than that of each single material but are lower than the sum of both. In contrast, when polarized in opposite directions, the electrostatic potential differences of the UN and NU configurations are greatly reduced. In addition, the UN and NU ferroelectric heterointerfaces display markedly weaker polarization-induced fields compared to their UU and NN counterparts. It demonstrates that there exists a certain charge transfer, which partially offsets the electrostatic potential difference. Under the manipulation of ferroelectric polarization, the net charge transfer is also substantial to the electrostatic potential difference at ferroelectric heterointerfaces, thereby mitigating the total ferroelectric polarization strength. Therefore, it is noted that the opposite UN and NU polarization configurations may manifest strikingly distinct characteristics in ferroelectric interlayer interactions, charge accumulation or separation, and carrier transport behaviors compared to the identical UU and NN configurations. Based on Figure 5a,d, it can be observed that the accumulation and dissipation of charges at the interface are approximately equal for both UU and NN configurations. This suggests that the redistribution of charges primarily occurs at the interface.

The phenomenon of spontaneous polarization arises in both α-In_2Se_3 and α-Ga_2Se_3 as a direct consequence of central symmetry breaking. It is attributed to the spatial displacement of positive and negative charges, leading to a non-uniform charge distribution within these materials. Upon the formation of a heterojunction, the charge redistribution occurs across the entire α-In_2Se_3/α-Ga_2Se_3 ferroelectric heterointerfaces, leading to the formation of an interfacial dipole moment. The presence of the interfacial dipole moment induces a built-in polarization field at the 2D ferroelectric interface, which in turn affects the energy band alignments and the motion behaviors of electrons and holes. Specifically, to analyze the interfacial charge transfer after α-In_2Se_3 and α-Ga_2Se_3 contact together, Figure 5 presents the charge density differences ($\Delta\rho(z)$) of 2D α-In_2Se_3/α-Ga_2Se_3 heterostructures in different stacking configurations. It is defined as

$$\Delta\rho(z) = \int \rho_{hetero}(x,y,z)dxdy - \int \rho_{In2Se3}(x,y,z)dxdy - \int \rho_{Ga2Se3}(x,y,z)dxdy,$$

where $\rho_{hetero}(x,y,z)$, $\rho_{In2Se3}(x,y,z)$, and $\rho_{Ga2Se3}(x,y,z)$ are the charge densities of α-In_2Se_3/α-Ga_2Se_3 heterostructures, α-In_2Se_3, and α-Ga_2Se_3 at the (x, y, z) position, respectively. The blue and red colors represent the charge accumulation and depletion, respectively. The ferroelectric polarization is directed from the negatively charged center to the positively charged center. In the UU configuration of Figure 5a, the negative polarization surface of α-In_2Se_3 is in contact with the positive polarization surface of α-Ga_2Se_3 at its 2D ferroelectric interface. And, the interfacial dipole interaction prompts the electrons and holes to separate apart from each other and accumulate at both ends of the UU heterojunction. It facilitates the rapid separation of photogenerated electron–hole pairs, causing the electrons in the α-Ga_2Se_3 layer to be repelled to the bottom surface and accumulate there. At the same time, the positive charge center of the α-Ga_2Se_3 layer is adjacent to the α-In_2Se_3 layer, causing the electrons of the α-In_2Se_3 layer to be attracted and accumulate near the interface. In contrast, due to the positive polarization surface of α-In_2Se_3 being adjacent to the negative polarization surface of α-Ga_2Se_3, the internal built-in polarization field drives the electrons and holes in the opposite direction, as shown in the NN configuration of Figure 5d, which are finally accumulated on the top and bottom surfaces of α-In_2Se_3 and α-Ga_2Se_3, respectively. On the other hand, both the UN and NU configurations demonstrate quite weak built-in polarization fields at 2D α-In_2Se_3/α-Ga_2Se_3 heterointerfaces. As depicted in Figure 5b, there exist the negatively polarized charges of both α-In_2Se_3 (P$\uparrow$) and α-Ga_2Se_3 (P$\downarrow$) at the interface of the UN configuration, which results in a hole potential well to capture the massive holes and accumulate there. According to Figure 4d, the electrostatic potential barrier also causes the electrons to be repelled from the interface of the UN configuration. Conversely, the interface of the NU configuration exhibits the positively polarized charges of both α-In_2Se_3 (P$\downarrow$) and α-Ga_2Se_3 (P$\uparrow$). As shown in Figure 5c, it is capable of capturing and accumulating a substantial amount of electrons at the interface. There presents an electron potential well as observed in Figure 4e. As a result, the switchable accumulation of electrons and holes has been successfully achieved at 2D α-In_2Se_3/α-Ga_2Se_3 heterointerfaces by manipulating nonvolatile ferroelectricity. This method provides valuable insights into the conversion of channel conductivity types from electrons to holes for 2D FeFETs.

As is well known, FeFETs offer low power consumption during writing/reading operations with high rapidity and high scalability that is compatible with CMOS technology. Note that the ferroelectric polarization has been experimentally demonstrated to be reversible by an electric field [18,19]. By utilizing the reversible out-of-plane ferroelectric polarization, 2D α-In_2Se_3/α-Ga_2Se_3 FeFETs have been proposed in Figure 6. By leveraging the charge transfer properties of the NU and UN configurations, a ferroelectric reconfigurable FeFET device model has been developed, where the top gate and bottom gates are used to independently control the polarization states of α-In_2Se_3 and α-Ga_2Se_3, and the source and drain terminals are positioned on both sides of the α-In_2Se_3/α-Ga_2Se_3 heterointerfaces. When the polarization electric field is applied, the polarization state of the ferroelectric layer exhibits ferroelectric hysteresis, which remains unchanged even after removing the electric field. For the NU configuration in Figure 6a, the channel carriers are electrons, indicating an n-channel FeFET. At $V_{gs} = 0$, the FeFET is conductive due to the presence of electrons in the channel. When increasing the negative bias voltage $V_{gs} < 0$, the n-channel starts to narrow, and ultimately enters the cut-off state once approaching the threshold voltage. Conversely, by reversing both the top and down gates, the ferroelectric polarization of 2D α-In_2Se_3/α-Ga_2Se_3 FeFETs is changed to the UN configuration. As shown in Figure 6b, the p-channel FeFET is in a conductive state due to the presence of holes in the channel at $V_{gs} = 0$. When further increasing the positive bias voltage $V_{gs} > 0$, the p-channel starts to narrow, gradually approaches the threshold voltage, and eventually enters the cut-off state. By switching the polarization states from NU to UN, it is worth mentioning that the FeFET channel current flows in opposite directions. As a result, the nonvolatile nature of ferroelectric polarization enables a stable transition between the n-channel and p-channel of 2D α-In_2Se_3/α-Ga_2Se_3 FeFETs, which helps to achieve fully complementary logic components. The ferroelectric modulation strategy provides significant guidance and opportunities for advanced data storage and processing technologies in nonvolatile electronic logic and memory applications.

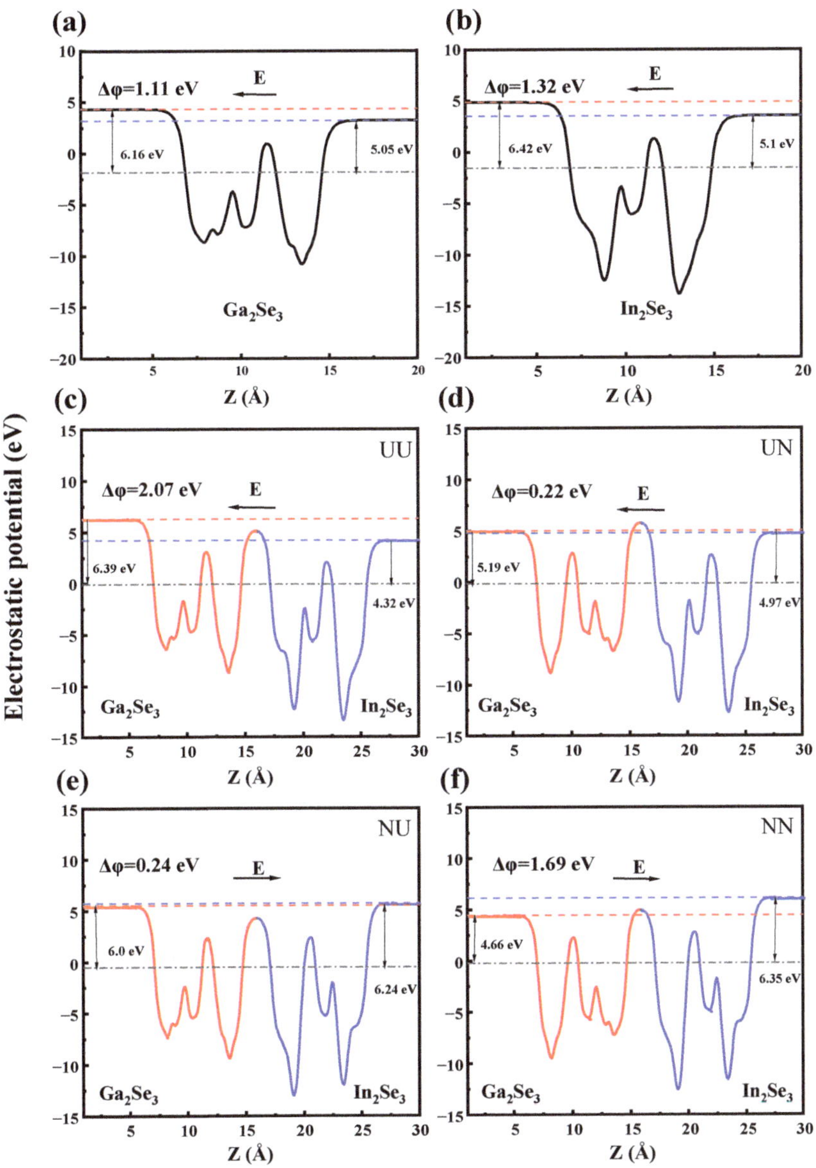

Figure 4. Planar average electrostatic potentials of (**a**) α-In_2Se_3, (**b**) α-Ga_2Se_3, (**c**) NN, (**d**) UN, (**e**) NU, and (**f**) NN configurations heterostructures along the z-direction, respectively. In (**c–f**), the blue and red parts of the electrostatic potential curves correspond to α-In_2Se_3 and α-Ga_2Se_3, respectively. The arrows indicate the directions of the polarization-induced electric fields. The vacuum levels and Fermi levels are marked with blue, red, and gray dashed lines, respectively.

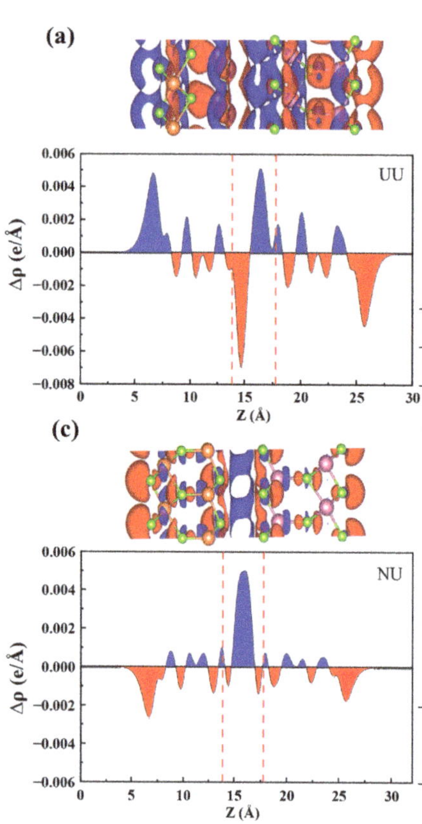

Figure 5. Plane-averaged charge density differences of the α-In_2Se_3/α-Ga_2Se_3 heterostructures in (**a**) UU, (**b**) UN, (**c**) NU, and (**d**) NN configurations along the z-direction, respectively. The red dashed lines represent the α-In_2Se_3/α-Ga_2Se_3 interface boundaries.

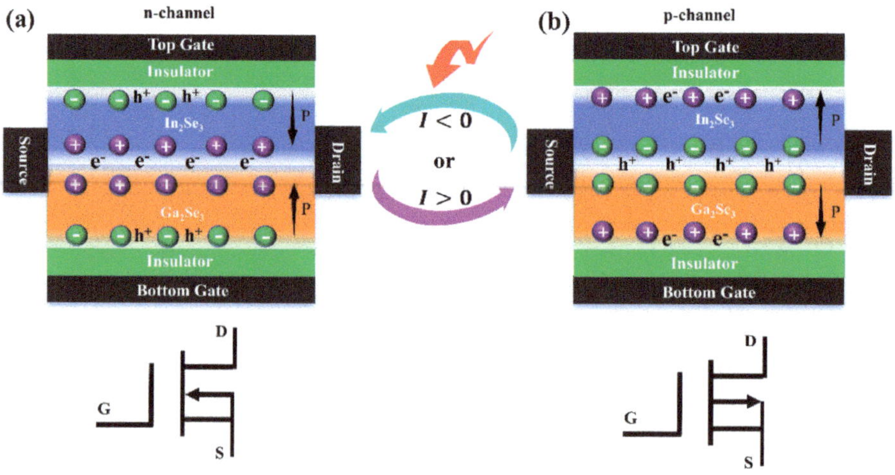

Figure 6. (**a**) The n-channel and (**b**) p-channel FeFET devices based on NU and UN configurations, respectively. The red arrow represents that the FeFET channel current flows in opposite directions. The "P" represents the ferroelectric polarization directions controlled by the top and bottom gates.

3.4. Strain Modulation on α-In_2Se_3/α-Ga_2Se_3 Polar Heterostructures

Owing to the exceptional mechanical flexibility and intrinsic piezoelectric characteristics of 2D ferroelectrics, strain engineering has emerged as a crucial technique for modulating their electronic band structures. In this study, we explored the modulation of the electronic characteristics of α-In_2Se_3/α-Ga_2Se_3 polar heterostructures by applying in-plane biaxial strains. Figure 7 shows the projected energy bands of α-In_2Se_3/α-Ga_2Se_3 polar heterostructures in UU, UN, NU, and NN configurations from −10% (compressive) to +10% (tensile), respectively. Because of belonging to the same III–VI family, it is noted that α-In_2Se_3 and α-Ga_2Se_3 have similar trends of electronic band structures on in-plane biaxial strains. As tensile strains increase, the CBMs of α-In_2Se_3 and α-Ga_2Se_3 decrease obviously while their VBMs gradually increase. It leads to decreasing bandgaps of both α-In_2Se_3 and α-Ga_2Se_3. In addition, the valence bands distinctly shift upward at the K point when applying tensile strains, leading to the VBM transition from the Γ point to the K point. On the other side, the bandgaps of α-In_2Se_3 and α-Ga_2Se_3 become narrower under high compressive strains. The substantial structural transformation results in changing the atomic interactions and making a distinct shift in charge distribution and band structures.

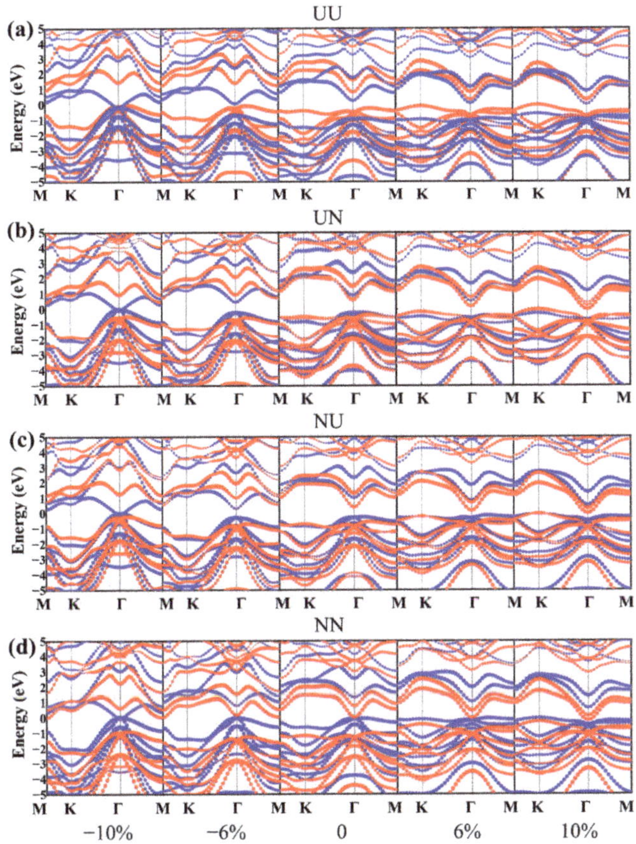

Figure 7. Projected band structures of (**a**) UU, (**b**) UN, (**c**) NU, and (**d**) NN heterostructures under biaxial strains of −10%, −6%, 0%, 6%, and 10%, in which the blue and red colors indicate the contributions of α-In_2Se_3 and α-Ga_2Se_3, respectively.

Furthermore, the band alignments of α-In_2Se_3/α-Ga_2Se_3 polar heterostructures are highly sensitive to the ferroelectric polarization configurations. When changing the configurations from UU to UN, NU, and NN in Figure 7, the electronic energy bands of α-Ga_2Se_3 and α-In_2Se_3 shift downward and upward significantly in opposite trends, respectively. As the compressive or tensile strain increases in Figure 8a,d, the CBM of α-In_2Se_3 becomes lower than the VBM of α-Ga_2Se_3 in the UU configuration, whereas the CBM of α-Ga_2Se_3 becomes lower than the VBM of α-In_2Se_3 in the NN configuration. As shown in Figure 8a,d, both the UU and NN configurations experience a band alignment transition spanning from type III to type II and back to type III. Conversely, in UN and NU configurations, the CBM and VBM of α-In_2Se_3 are positioned intermediately between those of α-Ga_2Se_3 under compressive strain, and the CBM and VBM of α-Ga_2Se_3 are situated between those of α-In_2Se_3 under tensile strain. Figure 8b,c demonstrate that both the UN and NU configurations undergo a band alignment transition covering type I→II→I. Consequently, strain can profoundly influence the shape of electronic band structures, modulate the bandgap values and their direct or indirect types, and tune the band alignment relationships in the α-In_2Se_3/α-Ga_2Se_3 ferroelectric heterostructures. The controllable transformation of energy bands facilitates the charge separation, transfer, recombination, and redistribution processes in the α-In_2Se_3/α-Ga_2Se_3 ferroelectric heterostructures. As a result, 2D III–VI ferroelectric junctions integrate electronic, optical, and memory capabilities for nonvolatile multifunctional applications.

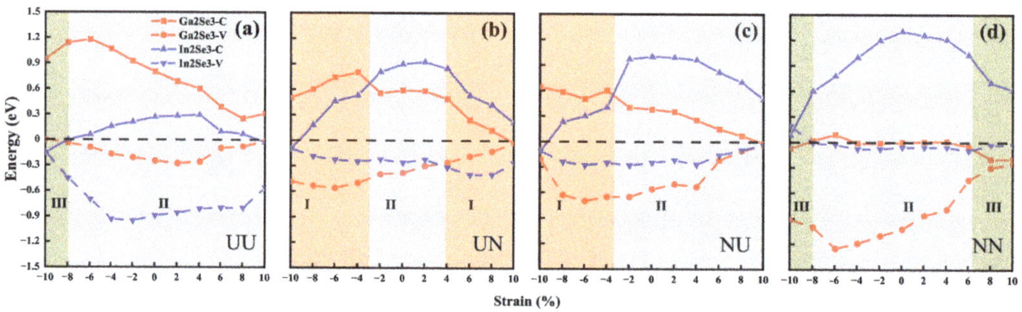

Figure 8. CBM and VBM values of α-In_2Se_3/α-Ga_2Se_3 heterostructures in (**a**) UU, (**b**) UN, (**c**) NU, and (**d**) NN configurations as a function of the in-plane biaxial strain, respectively. The red (blue) solid lines and dashed lines represent the CBMs and VBMs of α-Ga_2Se_3 (α-In_2Se_3), respectively. The orange, gray, and light green regions denote type-I, type-II, and type-III band structures, respectively.

4. Conclusions

In summary, the structural stability, electrostatic potential, interfacial charge transfer, and electronic band structures of 2D α-In_2Se_3/α-Ga_2Se_3 ferroelectric heterostructures have been thoroughly investigated by using first-principles calculations. The results indicate that α-In_2Se_3 has a stronger ferroelectric polarization compared to α-Ga_2Se_3. The polarization reversal of α-In_2Se_3 facilitates the nonvolatile manipulation of the built-in electric field within α-In_2Se_3/α-Ga_2Se_3 ferroelectric heterointerfaces, allowing for the reversible field directions. The pronounced built-in electric field in the UU and NN prompts the electrons and holes to separate apart from each other, leading to free carrier depletion. Conversely, the switchable accumulation of electrons and holes has been successfully achieved at the 2D ferroelectric interfaces of NU and UN. The nonvolatile nature of ferroelectric polarization enables a novel approach for modulating the electrical conductivity transition between the n-channel and p-channel in 2D α-In_2Se_3/α-Ga_2Se_3 FeFETs. Moreover, the band alignments of the α-In_2Se_3/α-Ga_2Se_3 ferroelectric heterostructures have been modulated by in-plane biaxial strains, achieving a type III–II–III transition in UU and NN, and a type I–II–I transition in UN and NU, respectively. Our findings demonstrate that 2D ferroelectric vdW heterostructures open promising avenues to utilize nonvolatile spontaneous polarization for the development of advanced data logic, optoelectronic memory, and computing techniques.

Author Contributions: Conceptualization, M.P.; Methodology, P.L., D.K., J.Y., S.C., Y.L., Q.C., Z.H., F.L., Y.X. and H.W.; Validation, P.L., D.K., J.Y., S.C., Y.L., Q.C., Z.H., F.L. and H.W.; Formal analysis, P.L., D.K., J.Y., S.C., Y.L., Q.C., Z.H., F.L. and Y.X.; Investigation, P.L., D.K., J.Y., S.C., Y.L., Q.C., Z.H., F.L., Y.X. and H.W.; Resources, Y.X., H.W., X.Z. and M.P.; Data curation, P.L., D.K., S.C. and Q.C.; Writing—original draft, P.L.; Writing—review & editing, M.P.; Supervision, X.Z. and M.P.; Project administration, M.P.; Funding acquisition, X.Z. and M.P. All authors have read and agreed to the published version of the manuscript.

Funding: This research was financially supported by the National Natural Science Foundation of China (grant numbers 62474018, 51402064), Beijing Natural Science Foundation (grant number 4222071), National Key R&D Program of China (grant number 2018YFA0703700), Fundamental Research Funds for the Central Universities (grant numbers FRF-BR-20-02A, FRF-TP-19-044A2, FRF-TP-20-016A2), Student Research Training Project (grant numbers 202421100A, 202421201A, 202021010), and Youth Innovation Promotion Association of the Chinese Academy of Sciences (grant number 2015387).

Data Availability Statement: The data that support the findings of this study are available from the corresponding author upon reasonable request.

Conflicts of Interest: The authors declare no conflicts of interest.

References

1. Wang, H.; Liu, Z.R.; Yoong, H.Y.; Paudel, T.R.; Xiao, J.X.; Guo, R.; Lin, W.N.; Yang, P.; Wang, J.; Chow, G.M.; et al. Direct observation of room-temperature out-of-plane ferroelectricity and tunneling electroresistance at the two-dimensional limit. *Nat. Commun.* **2018**, *9*, 3319. [CrossRef] [PubMed]
2. Morozovska, A.N.; Eliseev, E.A.; Glinchuk, M.D. Size effects and depolarization field influence on the phase diagrams of cylindrical ferroelectric nanoparticles. *Phys. B-Condens. Matter* **2006**, *387*, 358–366. [CrossRef]
3. Fong, D.D.; Stephenson, G.B.; Streiffer, S.K.; Eastman, J.A.; Auciello, O.; Fuoss, P.H.; Thompson, C. Ferroelectricity in Ultrathin Perovskite Films. *Science* **2004**, *304*, 1650–1653. [CrossRef]
4. Liu, F.; You, L.; Seyler, K.L.; Li, X.; Yu, P.; Lin, J.; Wang, X.; Zhou, J.; Wang, H.; He, H.; et al. Room-temperature ferroelectricity in $CuInP_2S_6$ ultrathin flakes. *Nat. Commun.* **2016**, *7*, 12357. [CrossRef] [PubMed]
5. Pang, K.; Xu, X.; Wei, Y.; Ying, T.; Li, W.; Yang, J.; Li, X.; Jiang, Y.; Zhang, G.; Tian, W. Integrating ferromagnetism and ferroelectricity in an iron chalcogenide monolayer: A first-principles study. *Nanoscale* **2022**, *14*, 14231–14239. [CrossRef] [PubMed]
6. Zhu, H.; Wang, Y.; Xiao, J.; Liu, M.; Xiong, S.; Wong, Z.J.; Ye, Z.; Ye, Y.; Yin, X.; Zhang, X. Observation of piezoelectricity in free-standing monolayer MoS_2. *Nat. Nanotechnol.* **2015**, *10*, 151–155. [CrossRef]
7. Ding, W.; Zhu, J.; Wang, Z.; Gao, Y.; Xiao, D.; Gu, Y.; Zhang, Z.; Zhu, W. Prediction of intrinsic two-dimensional ferroelectrics in In_2Se_3 and other III_2-VI_3 van der Waals materials. *Nat. Commun.* **2017**, *8*, 14956. [CrossRef]
8. Li, L.; Wu, M. Binary Compound Bilayer and Multilayer with Vertical Polarizations: Two-Dimensional Ferroelectrics, Multiferroics, and Nanogenerators. *ACS Nano* **2017**, *11*, 6382–6388. [CrossRef] [PubMed]
9. Wan, S.; Li, Y.; Li, W.; Mao, X.; Wang, C.; Chen, C.; Dong, J.; Nie, A.; Xiang, J.; Liu, Z.; et al. Nonvolatile Ferroelectric Memory Effect in Ultrathin α-In_2Se_3. *Adv. Funct. Mater.* **2019**, *29*, 1808606. [CrossRef]
10. Si, M.; Liao, P.-Y.; Qiu, G.; Duan, Y.; Ye, P.D. Ferroelectric Field-Effect Transistors Based on MoS_2 and $CuInP_2S_6$ Two-Dimensional van der Waals Heterostructure. *ACS Nano* **2018**, *12*, 6700–6705. [CrossRef] [PubMed]
11. Lu, H.; Lipatov, A.; Ryu, S.; Kim, D.J.; Lee, H.; Zhuravlev, M.Y.; Eom, C.B.; Tsymbal, E.Y.; Sinitskii, A.; Gruverman, A. Ferroelectric tunnel junctions with graphene electrodes. *Nat. Commun.* **2014**, *5*, 5518. [CrossRef]
12. Lu, J.; Yao, J.; Yan, J.; Gao, W.; Huang, L.; Zheng, Z.; Zhang, M.; Li, J. Strain engineering coupled with optical regulation towards a high-sensitivity In_2S_3 photodetector. *Mater. Horiz.* **2020**, *7*, 1427–1435. [CrossRef]
13. Gallego-Parra, S.; Vilaplana, R.; Gomis, O.; Rodríguez-Hernández, P.; Muñoz, A.; González, J.A.; Sans, J.A.; Popescu, C.; Manjón, F.J. High-Pressure Synthesis of β- and α-In_2Se_3-Like Structures in Ga_2S_3. *Chem. Mater.* **2022**, *34*, 6068–6086. [CrossRef]
14. Meng, S.; Wang, J.; Shi, H.; Sun, X.; Gao, B. Distinct ultrafast carrier dynamics of α-In_2Se_3 and β-In_2Se_3: Contributions from band filling and bandgap renormalization. *Phys. Chem. Chem. Phys.* **2021**, *23*, 24313–24318. [CrossRef]

15. Han, G.; Chen, Z.-G.; Drennan, J.; Zou, J. Indium Selenides: Structural Characteristics, Synthesis and Their Thermoelectri Performances. *Small* **2014**, *10*, 2747–2765. [CrossRef]
16. Chaiken, A.; Nauka, K.; Gibson, G.A.; Lee, H.; Yang, C.C.; Wu, J.; Ager, J.W.; Yu, K.M.; Walukiewicz, W. Structural and electron properties of amorphous and polycrystalline In_2Se_3 films. *J. Appl. Phys.* **2003**, *94*, 2390–2397. [CrossRef]
17. Li, W.; Sabino, F.P.; Crasto de Lima, F.; Wang, T.; Miwa, R.H.; Janotti, A. Large disparity between optical and fundamental ban gaps in layered. *Phys. Rev. B* **2018**, *98*, 165134. [CrossRef]
18. Zhou, Y.; Wu, D.; Zhu, Y.; Cho, Y.; He, Q.; Yang, X.; Herrera, K.; Chu, Z.; Han, Y.; Downer, M.C.; et al. Out-of-Plane Piezoelectricit and Ferroelectricity in Layered α-In_2Se_3 Nanoflakes. *Nano Lett.* **2017**, *17*, 5508–5513. [CrossRef]
19. Cui, C.; Hu, W.-J.; Yan, X.; Addiego, C.; Gao, W.; Wang, Y.; Wang, Z.; Li, L.; Cheng, Y.; Li, P.; et al. Intercorrelated In-Plan and Out-of-Plane Ferroelectricity in Ultrathin Two-Dimensional Layered Semiconductor In_2Se_3. *Nano Lett.* **2018**, *18*, 1253–125 [CrossRef]
20. Zhao, J.Z.; Chen, L.C.; Xu, B.; Zheng, B.B.; Fan, J.; Xu, H. Strain-tunable out-of-plane polarization in two-dimensional material *Phys. Rev. B* **2020**, *101*, 121407. [CrossRef]
21. Nukala, P.; Ahmadi, M.; Wei, Y.; de Graaf, S.; Stylianidis, E.; Chakrabortty, T.; Matzen, S.; Zandbergen, H.W.; Björling, A.; Manni D.; et al. Reversible oxygen migration and phase transitions in hafnia-based ferroelectric devices. *Science* **2021**, *372*, 630–63 [CrossRef] [PubMed]
22. Xue, F.; Zhang, J.; Hu, W.; Hsu, W.-T.; Han, A.; Leung, S.-F.; Huang, J.-K.; Wan, Y.; Liu, S.; Zhang, J.; et al. Multidirectio Piezoelectricity in Mono- and Multilayered Hexagonal α-In_2Se_3. *ACS Nano* **2018**, *12*, 4976–4983. [CrossRef] [PubMed]
23. Feng, W.; Gao, F.; Hu, Y.; Dai, M.; Li, H.; Wang, L.; Hu, P. High-performance and flexible photodetectors based on chemical vapo deposition grown two-dimensional In_2Se_3 nanosheets. *Nanotechnology* **2018**, *29*, 445205. [CrossRef]
24. Kong, D.; Tian, F.; Li, P.; Xu, Y.; Wei, H.; Zheng, X.; Peng, M. Manipulating Ferroelectric α-In_2Se_3/GaN Dipole Interactions b Polarization Engineering. *ACS Appl. Electron. Mater.* **2024**, *6*, 6747–6757. [CrossRef]
25. Si, M.; Zhang, Z.; Chang, S.-C.; Haratipour, N.; Zheng, D.; Li, J.; Avci, U.; Ye, P. Asymmetric Metal/α-In_2Se_3/Si Crossba Ferroelectric Semiconductor Junction. *ACS Nano* **2021**, *15*, 5689–5695. [CrossRef] [PubMed]
26. Shi, H.; Li, M.; Shaygan Nia, A.; Wang, M.; Park, S.; Zhang, Z.; Lohe, M.R.; Yang, S.; Feng, X. Ultrafast Electrochemical Synthesis of Defect-Free In_2Se_3 Flakes for Large-Area Optoelectronics. *Adv. Mater.* **2020**, *32*, 1907244. [CrossRef] [PubMed]
27. Zhang, G.; Lu, K.; Wang, Y.; Wang, H.; Chen, Q. Mechanical and electronic properties of $\alpha-M_2X_3$ (M = Ga, In; X = S, Se monolayers. *Phys. Rev. B* **2022**, *105*, 235303. [CrossRef]
28. Xue, W.; Jiang, Q.; Wang, F.; He, R.; Pang, R.; Yang, H.; Wang, P.; Yang, R.; Zhong, Z.; Zhai, T.; et al. Discovery of Robus Ferroelectricity in 2D Defective Semiconductor α-Ga_2Se_3. *Small* **2022**, *18*, 2105599. [CrossRef] [PubMed]
29. Liu, X.; Mao, Y. Controllable electrical contact characteristics of graphene/Ga_2X_3 (X = S, Se) ferroelectric heterojunctions. *Appl Phys. Lett.* **2024**, *125*, 043102. [CrossRef]
30. Hu, L.; Huang, X. Peculiar electronic, strong in-plane and out-of-plane second harmonic generation and piezoelectric properties of atom-thick α-M_2X_3 (M = Ga, In; X = S, Se): Role of spontaneous electric dipole orientations. *RSC Adv.* **2017**, *7*, 55034–55043 [CrossRef]
31. Wang, V.; Xu, N.; Liu, J.-C.; Tang, G.; Geng, W.-T. VASPKIT: A user-friendly interface facilitating high-throughput computing and analysis using VASP code. *Comput. Phys. Commun.* **2021**, *267*, 108033. [CrossRef]
32. Kresse, G.; Furthmüller, J. Efficient iterative schemes for ab initio total-energy calculations using a plane-wave basis set. *Phys. Rev. B* **1996**, *54*, 11169–11186. [CrossRef] [PubMed]
33. Sun, J.; Ruzsinszky, A.; Perdew, J.P. Strongly Constrained and Appropriately Normed Semilocal Density Functional. *Phys. Rev. Lett.* **2015**, *115*, 036402. [CrossRef]
34. Borlido, P.; Aull, T.; Huran, A.W.; Tran, F.; Marques, M.A.L.; Botti, S. Large-Scale Benchmark of Exchange-Correlation Functionals for the Determination of Electronic Band Gaps of Solids. *J. Chem. Theory Comput.* **2019**, *15*, 5069–5079. [CrossRef] [PubMed]
35. Sun, J.; Remsing, R.C.; Zhang, Y.; Sun, Z.; Ruzsinszky, A.; Peng, H.; Yang, Z.; Paul, A.; Waghmare, U.; Wu, X.; et al. Accurate first-principles structures and energies of diversely bonded systems from an efficient density functional. *Nat. Chem.* **2016**, *8*, 831–836. [CrossRef] [PubMed]
36. Nepal, N.K.; Ruzsinszky, A.; Bates, J.E. Rocksalt or cesium chloride: Investigating the relative stability of the cesium halide structures with random phase approximation based methods. *Phys. Rev. B* **2018**, *97*, 115140. [CrossRef]
37. Buda, I.G.; Lane, C.; Barbiellini, B.; Ruzsinszky, A.; Sun, J.; Bansil, A. Characterization of Thin Film Materials using SCAN meta-GGA, an Accurate Nonempirical Density Functional. *Sci. Rep.* **2017**, *7*, 44766. [CrossRef]
38. Blöchl, P.E. Projector augmented-wave method. *Phys. Rev. B* **1994**, *50*, 17953–17979. [CrossRef]
39. Kresse, G.; Joubert, D. From ultrasoft pseudopotentials to the projector augmented-wave method. *Phys. Rev. B* **1999**, *59*, 1758–1775. [CrossRef]
40. Grimme, S.; Antony, J.; Ehrlich, S.; Krieg, H. A consistent and accurate ab initio parametrization of density functional dispersion correction (DFT-D) for the 94 elements H-Pu. *J. Chem. Phys.* **2010**, *132*, 154104. [CrossRef] [PubMed]

41. Ayadi, T.; Debbichi, L.; Badawi, M.; Said, M.; Rocca, D.; Lebègue, S. An ab initio study of the electronic properties of the ferroelectric heterostructure In_2Se_3/Bi_2Se_3. *Appl. Surf. Sci.* **2021**, *538*, 148066. [CrossRef]
42. Xiong, W.; Xia, C.; Zhao, X.; Wang, T.; Jia, Y. Effects of strain and electric field on electronic structures and Schottky barrier in graphene and SnS hybrid heterostructures. *Carbon* **2016**, *109*, 737–746. [CrossRef]
43. Li, X.; Zhai, B.; Song, X.; Yan, Y.; Li, J.; Xia, C. Two-dimensional Janus-In_2STe/InSe heterostructure with direct gap and staggered band alignment. *Appl. Surf. Sci.* **2020**, *509*, 145317. [CrossRef]

nanomaterials

Communication

SnS_2 as a Saturable Absorber for Mid-Infrared Q-Switched Er:SrF_2 Laser

Chun Li [1], Qi Yang [1,*], Yuqian Zu [1], Syed Zaheer Ud Din [1], Yu Yue [2], Ruizhan Zhai [1] and Zhongqing Jia [1]

1 International School for Optoelectronic Engineering, Qilu University of Technology (Shandong Academy of Sciences), Jinan 250353, China; lichun@qlu.edu.cn (C.L.); zuyuqian@qlu.edu.cn (Y.Z.); zaheer@qlu.edu.cn (S.Z.U.D.); zrz@vip.sdlaser.cn (R.Z.); jiazhongqing@vip.sdlaser.cn (Z.J.)

2 School of Science, Shandong Jianzhu University, Jinan 250101, China; yueyu19@sdjzu.edu.cn

* Correspondence: shanshiyangqi@126.com or qiyang@qlu.edu.cn

Abstract: Two-dimensional (2D) materials own unique band structures and excellent optoelectronic properties and have attracted wide attention in photonics. Tin disulfide (SnS_2), a member of group IV-VI transition metal dichalcogenides (TMDs), possesses good environmental optimization, oxidation resistance, and thermal stability, making it more competitive in application. By using the intensity-dependent transmission experiment, the saturable absorption properties of the SnS_2 nanosheet nearly at 3 μm waveband were characterized by a high modulation depth of 32.26%. Therefore, a few-layer SnS_2 was used as a saturable absorber (SA) for a bulk Er:SrF_2 laser to research its optical properties. When the average output power was 140 mW, the passively Q-switched laser achieved the shortest pulse width at 480 ns, the optimal single pulse energy at 3.78 μJ, and the highest peak power at 7.88 W. The results of the passively Q-switched laser revealed that few-layer SnS_2 had an admirable non-linear optical response at near 3 μm mid-infrared solid-state laser.

Keywords: 2D SnS_2; TMDs; non-linear optical materials; mid-infrared laser

1. Introduction

Mid-infrared lasers have numerous practical applications due to their ability to cover multiple atmospheric windows, their capacity for strong absorption of a variety of molecules, and their ability to effectively concentrate thermal radiation energy [1,2]. Particularly, lasers with 3 μm wavelengths possess a superior capability when it comes to water absorption compared to other mid-infrared lasers [3]. K. S. Bagdasarov et al. produced the first 2.94 μm laser output at room temperature using an Er:YAG crystal in 1983, which paved the way for the study of the 3 μm laser [4]. According to research, highly doped Er ions are necessary to generate an efficient laser output when YAG-like oxides are used as the laser hosts [5,6]. However, it is challenging to generate crystals with high doping concentrations due to the limitations of the growing techniques. Fortunately, Er:SrF_2 crystals, with low doping concentrations, have a high specific capacity to generate a 3 μm laser, which could avoid the difficulty of highly doping Er:YAG materials. In this study, bulk Er:SrF_2 crystals were adopted as the gain material which was grown by the simple and cost-effective temperature gradient technique (TGT). And the doping concentration of Er ions was only 3%. Meanwhile, Er:SrF_2 crystals, as the gain material, have excellent optical, mechanical, and thermal properties [6,7].

Passively Q-switched lasers could generate high-energy pulses up to several millijoules with a simple laser structure. They gained significant applications in scientific research and medical treatment. Recently, researchers have shown that two-dimensional (2D) materials have the capacity to act as excellent saturable absorber (SA) materials, which further enhances the potential development and application prospects of pulse lasers [8–12]. Owing to their strong interaction with light, relatively high-charge carrier mobilities, exotic electronic properties, and excellent mechanical characteristics, transition metal dichalcogenides (TMDs) have attracted in-depth investigation and developed extraordinary applications [13–15]. Especially in the laser field, TMDs commonly possess a large modulation depth which is advantageous for great Q-switched pulse lasers [16,17]. Until now, many kinds of 2D TMDs, such as MoS_2, SnS_2, $SnSe_2$, ReS_2, and $MoTe_2$, have been developed to apply in pulse lasers as SA materials [17–22]. Among them, Tin disulfide (SnS_2) belongs to the IV-VI group TMDs with the CdI_2 crystal structure [21]. Two layers are combined through van der Waals forces, which makes it easy to prepare 2D structures. Furthermore, 2D SnS_2 owns excellent optical and electrical properties, making it widely used in the fields of ultrafast photonics and lasers [22–24]. Moreover, SnS_2 consists of Sn and S elements which are abundantly stored elements in nature, and they own excellent properties of environmental optimization, low-cost, and nontoxicity [25]. Furthermore, SnS_2 has oxidation resistance and thermal stability, which is beneficial to improve the application stability of devices made of SnS_2 materials [26].

Research has shown that SnS_2 was a direct bandgap semiconductor having a value of 2.24 eV [27]. Based on the Planck formula, the bandgap determines the absorbed photons by SnS_2 in the visible optical energy region [27]. However, SnS_2 SA has been achieved in near-infrared lasers, which mainly focus on the Yb-doped and Er-doped fiber laser in the wavelength range of 1–2 μm [25–28]. Compared with fiber lasers, solid-state lasers have the advantages of high power, simple structure, and excellent efficiency. So far, whether 2D SnS_2 could be used as a SA in 3 μm solid-state lasers has not been verified yet.

In this study, the passively Q-switched Er:SrF_2 lasers have been investigated with the few-layer SnS_2 nanosheets employed as SA, which possess near 3 μm saturable absorption characteristics. Based on few-layer SnS_2 nanosheets, the Er:SrF_2 pulse laser was investigated using three kinds of resonators. When the output mirror had a radius of 100 mm and transmission of 1%, the shortest pulse width was 480 ns with a repetition rate of 37 kHz. Under the maximum absorbed power of 2.87 W, the laser acquired a single pulse energy of 3.78 μJ and a peak power of 7.88 W. At the output mirror of a radius of 100 mm and transmission of 4%, the Q-switched laser obtained the shortest pulse width of 820 ns, the maximum average output power of 87 mW, the single pulse energy of 2.18 μJ, and the peak power of 2.65 W. While the output mirror had a radius of 200 mm and transmission of 1%, the maximum average output power, repetition rate, pulse width, single pulse energy, and peak power were 149 mW, 40 kHz, 760 ns, 3.73 μJ, and 4.90 W, respectively.

2. Characterization of SnS_2 SA

The morphologies were observed with scanning electron microscopy (SEM) and transmission electron microscopy (TEM), shown in Figure 1. Figure 1a depicts the SEM image assembled by a large number of thick nanosheets. Therefore, SnS_2 needs to be prepared into a thin layer structure. A total of 50 mg of SnS_2 powder was added to a 10 mL centrifuge tube filled with alcohol. Through 24 h ultrasonic and 15 min centrifugal treatment, the resulting supernatant was dropped onto a YAG-substrate with a diameter of 12.7 mm and left to air dry for 12 h. After the ultrasonic exfoliation method, the few-layer nanosheet structures were confirmed, as shown in Figure 1b.

The Raman spectrum was characterized for the SnS_2 sample in Figure 2a. The peaks that are highlighted at 316.65 cm^{-1} and 206.04 cm^{-1} are attributed to the A_{1g} and E_g intralayer modes [28]. The absorption spectrum is measured when the wavelength is changed from 1.8 μm to 3 μm. As seen in Figure 2b, the SnS_2 nanosheets show a relatively low, flat, and broad absorption, which indicates that SnS_2 nanosheets are a promising broadband optical SA. The high transmission, in the mid-infrared band, is approximately 80–85%.

The non-linear saturable absorption was characterized in an intensity-dependent transmission experiment using a homemade laser with a repetition rate of 1 kHz and a pulse width of 1 μs at the wavelength of about 2.7 μm. Based on the Formula (1),

$$T(I) = 1 - T_{ns} - \Delta T * exp(-I/I_{sat}) \quad (1)$$

where $T(I)$: transmission; I: input intensity of the 2.7 μm laser; T_{ns}: non-saturable absorbance; ΔT: modulation depth; I_{sat}: saturation intensity. The experimental data and the fitted function are shown in Figure 3. The non-saturable absorbance was 23.07%, and the saturation intensity was 0.56 mJ/mm^2. Moreover, the modulation depth was high, up to 32.26%, which indicates that the SnS_2 materials have the capability as SA for the 2.7 μm passively Q-switched laser.

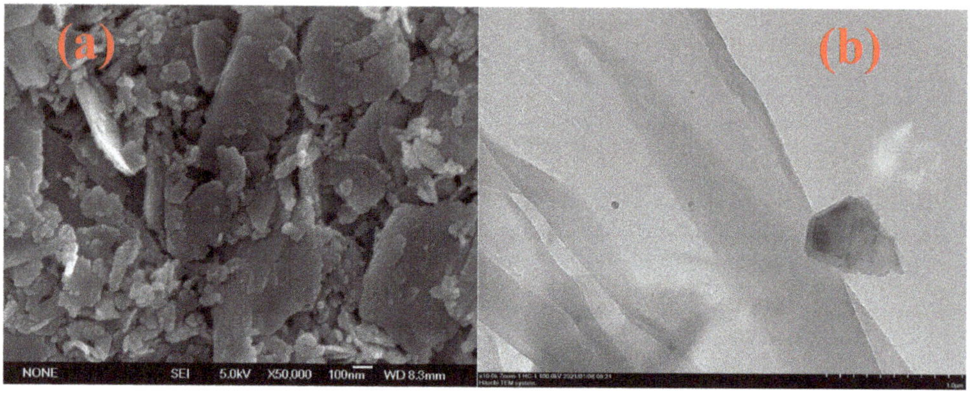

Figure 1. (**a**) SEM and (**b**) TEM images of the SnS_2 sample.

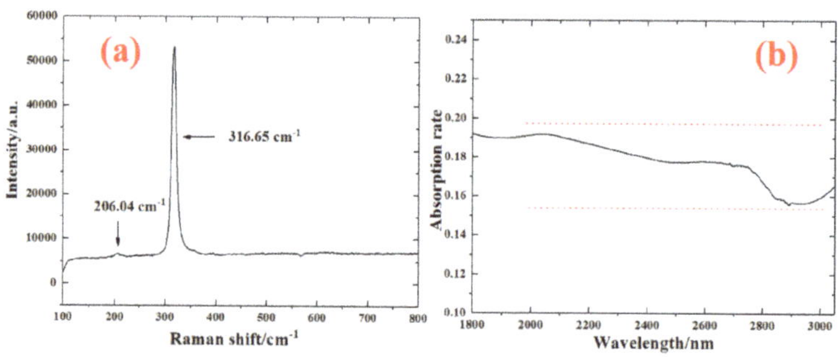

Figure 2. (**a**) The Raman spectrum and (**b**) absorption rate for the SnS_2 sample.

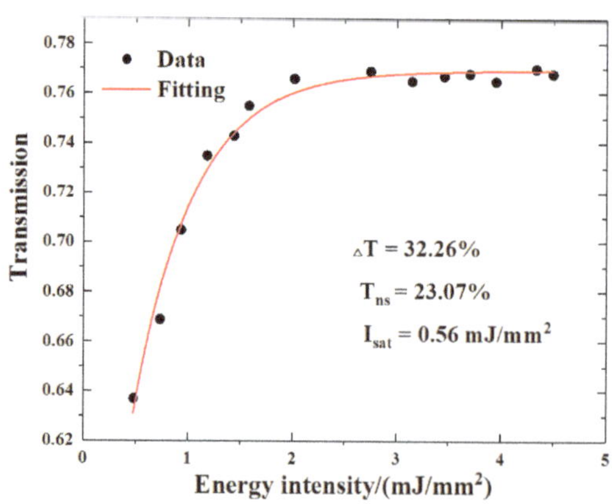

Figure 3. The function between transmission and energy intensity of SnS_2 SA.

3. Pulse Laser Experiments

The Er:SrF_2 passively Q-switched laser experiments were pumped by a laser diode (LD), emitting a continuous wave (CW) at a central wavelength of 976 nm. The pump laser was coupled by a fiber with a 105 µm radius. Using an optics system of 1:1, the laser was converged to the Er:SrF_2 crystal with dimensions 3 mm × 3 mm × 10 mm. The Er:SrF_2 crystal was mounted on a Cu holder. And the Cu holder was cooled with water to remove the excess heat and thus reduce the thermal effect. Mirrors M1 and M2 formed the resonator of the laser (shown in Figure 4). M1, working as an input mirror, was a plane mirror. M2 was the output mirror with a different radius and transmission. Three pulse laser experiments were investigated, when M2 had a radius of 100 mm with a transmission of 1% (T = 1%, R = 100 mm), a radius of 100 mm with a transmission of 4% (T = 4%, R = 100 mm), and a radius of 200 mm with a transmission 1% (T = 1%, R = 200 mm), respectively. When M2 had a radius of 100 mm, the length between M1 and M2 was the same as 80 mm. The Er:SrF_2 laser was first operated as a CW laser; then, the SnS_2 sample was placed into the resonant cavity. After carefully adjusting M1, M2, and the SnS_2 sample, the passively Q-switched lasers were constructed. At this time, the SnS_2 SA was 32 mm away from M1, and the spot size on the SA was calculated to be about 145 µm. While the radius of M2 was 200 mm, the cavity length was changed to 180 mm. When SnS_2 SA was located 40 mm away from M1 with a spot size of approximately 170 µm, the stable Q-switched laser was obtained.

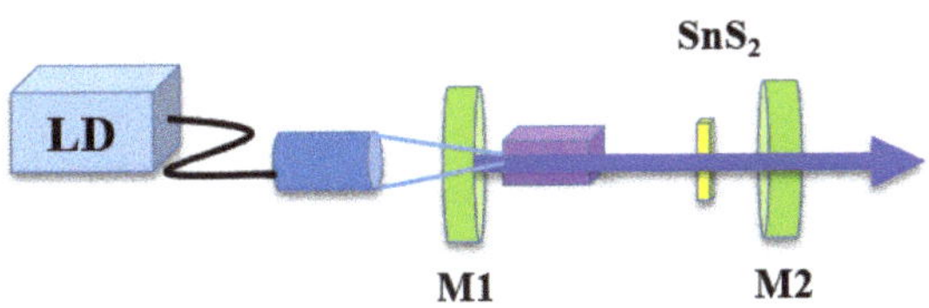

Figure 4. Experimental design of Er:SrF_2 passively Q-switched laser.

After the SnS_2 SA was employed in the laser cavity, the Er:SrF_2 passively Q-switched laser was successfully set up. For three laser cavities (T = 1%, R = 100 mm; T = 4%, R = 100 mm; T = 1%, R = 200 mm), the average output power and the absorbed pump power, shown in Figure 5a, all had a linear relationship. And the slope efficiencies were 5.22%, 4.28%, and 5.22%, respectively. When the absorbed pump power was increased to 2.87 W, the maximum average output power was 140 mW, 87 mW, and 149 mW, respectively. Comparing different transmittances, the average output power and slope efficiency acquired at a transmittance of 1% were higher than those of 4%, and the threshold absorbed pump power was lower because higher transmittance caused more loss. And, as shown in Figure 5b, when the transmittance was 1%, the passively Q-switched lasers were both dual-wavelength, located at 2729 nm and 2747 nm. At a transmittance of 4%, the central wavelength was 2728 nm.

As the absorbed pump power was increased from 0.53 W to 2.87 W, the pulse repetition rate gradually increased. To the three Q-switched lasers ((T = 1%, R = 100 mm), (T = 4%, R = 100 mm), and (T = 1%, R = 200 mm)), the highest pulse repetition rates were 37 kHz, 40 kHz, and 40 kHz, respectively. The detailed change rule is shown in Figure 6a. As shown in Figure 6b, the pulse widths were reduced with the increase of absorbed pump power. Three Q-switched lasers obtained the minimum pulse widths of 480 ns, 820 ns, and 760 ns. Comparing the results, the difference in repetition rate was small. Under the same transmission of 1%, the compact cavity design contributed to the compression pulse width.

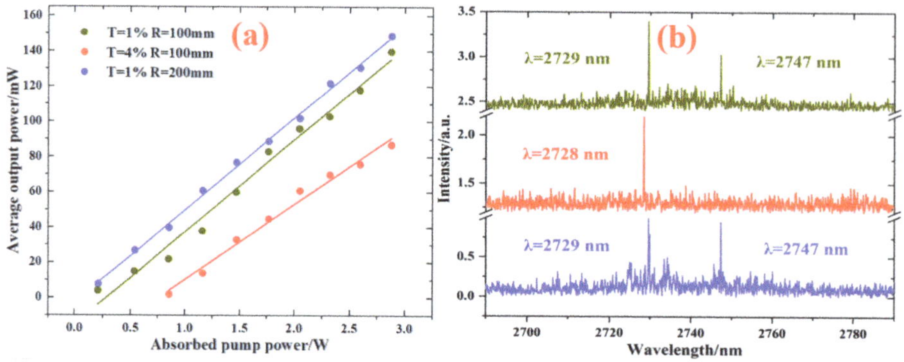

Figure 5. (**a**) Average output power versus absorbed pump power and (**b**) the Q-switched spectra.

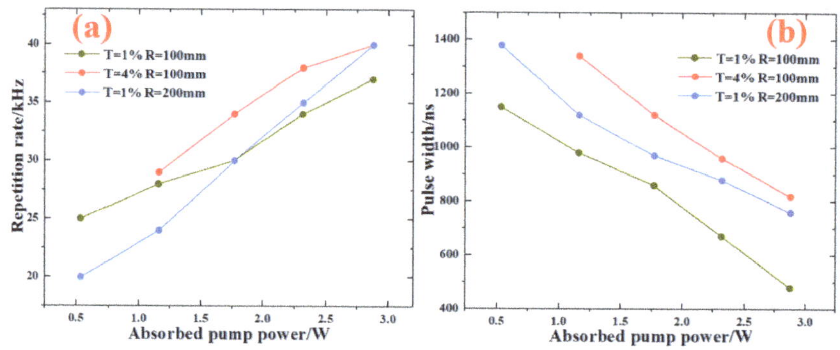

Figure 6. Repetition rate (**a**) and pulse width (**b**) as a function of absorbed pump power.

Under the absorbed pump power of 2.87 W, the pulse shape, using three laser cavities, is shown in Figure 7. As can be seen, the pulse train exhibits fine repeatability. The single pulse has a good Q-switched waveform at 480 ns, 820 ns, and 760 ns, respectively. Comparing the results of pulse width, the short cavity length and low transmittance result in higher intracavity power density, effectively compressing laser pulse width. When Er:SrF_2 laser absorbed the pump power of 2.87 W, the experiment acquired the single pulse energy of 3.78 μJ, 2.18 μJ, and 3.73 μJ. And the peak power was 7.88 W, 2.65 W, and 4.90 W, respectively. To sum up, when the radius of M2 was 100 mm with a transmission of 1%, the Er:SrF_2 passively Q-switched laser obtained superb results.

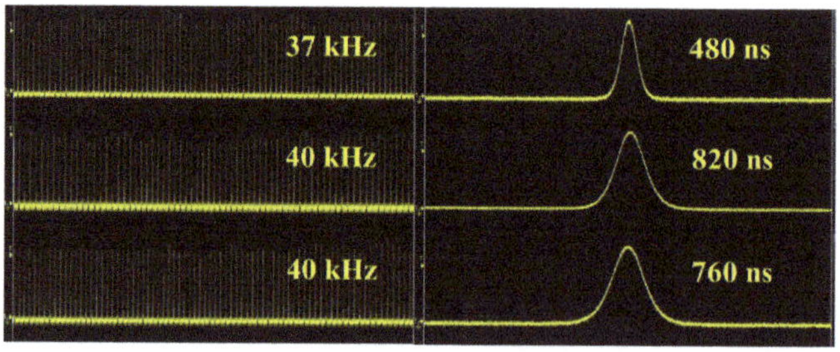

Figure 7. Pulse train and a single pulse of Er:SrF_2 pulse laser at an absorbed pump power of 2.87 W.

Table 1 lists the data of some experiments using 2D SA materials for 2.7–3 μm pulse lasers in recent years. Owing to the exotic photoelectric properties, many 2D materials, working as SAs, have been researched for mid-infrared lasers, such as graphene, Bi_2Se_3, MoS_2, BP, and so on. Due to its wide variety, 2D TMDs materials have received widespread attention, like $TiSe_2$, WSe_2, and ReS_2. In this paper, the SnS_2, belonging to TMDs, has proved a suitable SA for all-solid-state 2.7 μm lasers, having promoted the development of mid-infrared lasers. Moreover, compared with the other 2D SAs, SnS_2 owns a modulation depth of up to 32.26%, making it an excellent SA material for Q-switched lasers.

Table 1. Experimental results of bulk Q-switched lasers at near 3 μm by different 2D SAs.

SA	Gain	Modulation Depth /%	Output Power /mW	Pulse Width /ns	Repetition Rate /kHz	Pulse Energy /μJ	Wavelength/μm	Ref.
$TiSe_2$	Ho,Pr:LLF	9.2	130	160.5	98.8	1.32	2.95	[29]
BP	$Er:Lu_2O_3$	8	755	359	107	7.1	2.84	[30]
TiC	$Er:Lu_2O_3$	5.5	896	266.8	136.9	6.5	2.85	[31]
WSe_2	$Er:Lu_2O_3$	5.3	776	280	121	6.6	2.85	[32]
Nb_2CT_x	$Er:Lu_2O_3$	7.1	542	223.7	142.9	3.79	2.85	[33]
$Ti_4N_3T_x$	$Er:Lu_2O_3$	8.6	778	278.4	113.7	6.84	2.85	[34]
MoS_2	$Er:Lu_2O_3$	20.7	1030	335	121	8.5	2.84	[35]
$ReSe_2$	Er:YAP	7.5	526	202.8	244.6	2.2	2.73 + 2.80	[36]
ReS_2	$Er:SrF_2$	3.8	580	508	49	12.1	2.79	[37]
Bi	$Er:SrF_2$	1.82	226	980	56.2	4.02	2.73 + 2.75	[38]
SnS_2	$Er:SrF_2$	32.26	140	480	37	3.78	2.73 + 2.75	This work
SnS_2	$Er:SrF_2$	32.26	87	820	40	2.18	2.73	This work
SnS_2	$Er:SrF_2$	32.26	149	760	40	3.73	2.73 + 2.75	This work

4. Conclusions

Through experimental research, 2D SnS_2 has been proven to have exceptional saturable absorption characteristics in near 3 μm mid-infrared laser. According to the intensity-dependent non-linear optical absorption theory, non-saturable absorbance, modulation depth, and saturation intensity were 23.07%, 32.26%, and 0.56 mJ/mm^2, respectively. The diode-pumped $Er:SrF_2$ laser adopted a compact plane-concave cavity. The laser operation realized the highest pulse energy of 3.78 μJ and a pulse peak power of 7.88 W, when the maximum average output power was 140 mW, and the shortest pulse duration was 480 ns at a repetition rate of 37 kHz, under M2 with a radius of 100 mm and transmission of 1%. Employing M2 with a radius of 100 mm and transmission of 4%, the Q-switched laser obtained the shortest pulse width of 820 ns, single pulse energy of 2.18 μJ, and peak power of 2.65 W. Using an M2 with a radius of 200 mm and transmission of 1%, the maximum average output power, the single pulse energy, and the peak power of the $Er:SrF_2$ pulse laser were 149 mW, 3.73 μJ, and 4.90 W. The experiment results demonstrated that SnS_2, having a high modulation depth, could act as a SA of solid-state laser with nearly 3 μm, which improves the selectivity of mid-infrared SA.

Author Contributions: Investigation, C.L.; data curation, Q.Y. and Y.Z.; writing—review and editing, S.Z.U.D. and Y.Y.; supervision, R.Z. and Z.J. All authors have read and agreed to the published version of the manuscript.

Funding: The authors acknowledge support from the Natural Science Foundation of Shandong Province of China (ZR2021QF128, ZR2022QF060, ZR2020QA074); National Natural Science Foundation of China (12004208, 62105186); Qilu University of Technology (Shandong Academy of Sciences), Education and Industry Integration and Innovation Pilot (2022PY022); Key R&D Program of Shandong Province (2021CXGC010202); Major innovation projects for integrating science, education & industry of Qilu University of Technology (Shandong Academy of Sciences) (2022JBZ01-04).

Data Availability Statement: The data presented in this study are available upon request from the corresponding author.

Conflicts of Interest: The authors declare no conflict of interest.

References

1. Seddon, A.B. A prospective for new mid-infrared medical endoscopy using chalcogenide glasses. *Int. J. Appl. Glass Sci.* **2011**, 177–191. [CrossRef]
2. Xiong, Z.; Jiang, L.; Cheng, T.; Jiang, H. 100 Hz repetition-rate 2.794 μm Cr,Er:YSGG passively Q-switched laser with Fe^{2+}:ZnS saturable absorber. *Infrared Phys. Technol.* **2022**, *122*, 104087. [CrossRef]
3. Serebryakov, V.A.; Boĭko, É.V.; Petrishchev, N.N.; Yan, A.V. Medical applications of mid-IR lasers-problems and prospects. *J. Opt. Technol.* **2010**, *77*, 6–17. [CrossRef]
4. Bagdasarov, K.S.; Zhekov, V.I.; Lobachev, V.A.; Murina, T.M.; Prokhorov, A.M. Steady-state emission from a $Y_3Al_5O_{12}$: Er^{3+} laser (λ = 2.94 μ, T = 300° K). *Sov. J. Quantum Electron.* **1983**, *13*, 262–263. [CrossRef]
5. Ren, X.; Wang, Y.; Zhang, J.; Tang, D.; Shen, D. Short-pulse-width repetitively Q-switched ~2.7-μm Er:Y_2O_3 ceramic laser. *Appl. Sci.* **2017**, *7*, 1201. [CrossRef]
6. Ma, W.; Qian, X.; Wang, J.; Liu, J.; Fan, X.; Liu, J.; Su, L.; Xu, J. Highly efficient dual-wavelength mid-infrared CW laser in diode end-pumped Er:SrF_2 single crystals. *Sci. Rep.* **2016**, *6*, 36635. [CrossRef]
7. Liu, J.; Liu, J.; Guo, Z.; Zhang, H.; Ma, W.; Wang, J.; Su, L. Dual-wavelength Q-switched Er:SrF_2 laser with a black phosphorus absorber in the midinfrared region. *Opt. Express* **2016**, *24*, 30289–30295. [CrossRef]
8. Wang, Y.; Wang, J.; Wen, Q. MXene/Graphene oxide heterojunction as a saturable absorber for passively Q-switched solid-state pulse lasers. *Nanomaterials* **2021**, *11*, 720. [CrossRef]
9. Zhang, Y.; Lu, D.; Yu, H.; Zhang, H. Low-dimensional saturable absorbers in the visible spectral region, Adv. *Opt. Mater.* **2018**, 1800886. [CrossRef]
10. Guo, X.; Wang, S.; Yan, P.; Wang, J.; Yu, L.; Liu, W.; Zheng, Z.; Guo, C.; Ruan, S. High modulation depth enabled by $Mo_2Ti_2C_3T_x$ MXene for Q-switched pulse generation in a mid-infrared fiber laser. *Nanomaterials* **2022**, *12*, 1343. [CrossRef]
11. Liu, J.; Yang, F.; Lu, J.; Ye, S.; Guo, H.; Nie, H.; Zhang, J.; He, J.; Zhang, B.; Ni, Z. High output mode-locked laser empowered by defect regulation in 2D Bi_2O_2Se saturable absorber. *Nat. Commun.* **2022**, *13*, 3855. [CrossRef] [PubMed]
12. Feng, T.; Mao, D.; Cui, X.; Li, M.; Song, K.; Jiang, B.; Lu, H.; Quan, W. A filmy black-phosphorus polyimide saturable absorber for Q-switched operation in an erbium-doped fiber laser. *Materials* **2016**, *9*, 917. [CrossRef] [PubMed]
13. Mak, K.F.; Shan, J. Photonics and optoelectronics of 2D semiconductor transition metal dichalcogenides. *Nat. Photonics* **2016**, *10*, 216–226. [CrossRef]
14. Yang, Z.; Yang, Q.; Ren, X.; Tian, Y.; Zu, Y.; Li, C.; Din, S.Z.U.; Leng, J.; Liu, J. Passively mode-locked red Pr:$LiYF_4$ laser based on a two-dimensional palladium diselenide saturable absorber. *Opt. Express* **2022**, *30*, 2900–2908. [CrossRef] [PubMed]
15. Fadhel, M.M.; Ali, N.; Rashid, H.; Sapiee, N.M.; Hamzah, A.E.; Zan, M.S.D.; Aziz, N.A.; Arsad, N. A review on rhenium disulfide: Synthesis approaches, optical properties, and applications in pulsed lasers. *Nanomaterials* **2021**, *11*, 2367. [CrossRef]
16. Zhang, B.; Liu, J.; Wang, C.; Yang, K.; Lee, C.; Zhang, H.; He, J. Recent progress in 2D material-based saturable absorbers for all solid-state pulsed bulk lasers. *Laser Photonics Rev.* **2020**, *14*, 1900240. [CrossRef]
17. Li, C.; Yang, Z.; Yang, Q.; Zu, Y.; Din, S.Z.U.; Li, H.; Li, M. LD pumped passively Q-switched Pr:YLF lasers using VS_2 SA. *Opt Mater. Express* **2022**, *12*, 4191–4198. [CrossRef]
18. Su, X.; Zhang, B.; Wang, Y.; He, G.; Li, G.; Lin, N.; Yang, K.; He, J.; Liu, S. Broadband rhenium disulfide optical modulator for solid-state lasers. *Photonics Res.* **2018**, *6*, 498–505. [CrossRef]
19. Yan, B.; Zhang, B.; Nie, H.; Wang, H.; Li, G.; Sun, X.; Wang, R.; Lin, N.; He, J. High-power passively Q-switched 2.0 μm all-solid-state laser based on a $MoTe_2$ saturable absorber. *Opt. Express* **2018**, *26*, 18505–18512. [CrossRef]
20. Liu, X.; Li, X.; Lv, S.; Luo, W.; Xu, W.; Shi, Z.; Ren, Y.; Zhang, C.; Zhang, K. Electrochemical peeling few-layer $SnSe_2$ for high-performance ultrafast photonics. *ACS Appl. Mater. Interfaces* **2020**, *12*, 43049–43057.
21. Niu, K.; Chen, Q.; Sun, R.; Man, B.; Zhang, H. Passively Q-switched erbium-doped fiber laser based on SnS_2 saturable absorber. *Opt. Mater. Express* **2017**, *7*, 3934–3943. [CrossRef]
22. Shi, Z.; Sun, X.; Xie, W.; Chang, P.; Li, S.; Zhang, L.; Yang, X. Passively Q-switched Tm:YAP laser based on SnS_2 saturable absorber. *Optik* **2022**, *264*, 169421. [CrossRef]
23. Feng, T.; Zhang, D.; Li, X.; Abdul, Q.; Shi, Z.; Lu, J.; Guo, P.; Zhang, Y.; Liu, J.; Wang, Q.J. SnS_2 nanosheets for Er-doped fiber lasers. *ACS Appl. Nano Mater.* **2019**, *3*, 674–681. [CrossRef]
24. Liu, G.; Lyu, Y.; Li, Z.; Wu, T.; Yuan, J.; Yue, X.; Zhang, H.; Zhang, F.; Fu, S. Q-switched erbium-doped fiber laser based on silicon nanosheets as saturable absorber. *Optik* **2020**, *202*, 163692. [CrossRef]
25. Liu, M.; Wu, H.; Liu, X.; Wang, Y.; Lei, M.; Liu, W.; Guo, W.; Wei, Z. Optical properties and applications of SnS_2 SAs with different thickness. *Opto-Electronic Adv.* **2021**, *4*, 200029. [CrossRef]
26. Niu, K.; Sun, R.; Chen, Q.; Man, B.; Zhang, H. Passively mode-locked Er-doped fiber laser based on SnS_2 nanosheets as a saturable absorber. *Photonics Res.* **2018**, *6*, 72–76. [CrossRef]
27. Li, J.; Zhao, Y.; Chen, Q.; Niu, K.; Sun, R.; Zhang, H. Passively mode-locked ytterbium-doped fiber laser based on SnS_2 as saturable absorber. *IEEE Photonics J.* **2017**, *9*, 1506707. [CrossRef]
28. Gao, J.; Pan, J.; Liu, Y.; Guo, Q.; Han, X.; Shang, X.; Guo, L.; Zuo, Z.; Man, B.; Zhang, H.; et al. Observation of the dispersion effect of SnS_2 nanosheets in all-normaldispersion Yb-doped mode-locked fiber laser. *Infrared Phys. Technol.* **2019**, *102*, 102982. [CrossRef]

29. Nie, H.; Sun, X.; Zhang, B.; Yan, B.; Li, G.; Wang, Y.; Liu, J.; Shi, B.; Liu, S.; He, J. Few-layer $TiSe_2$ as a saturable absorber for nanosecond pulse generation in 2.95 μm bulk laser. *Opt. Lett.* **2018**, *43*, 3349–3352. [CrossRef]
30. Fan, M.; Li, T.; Zhao, S.; Li, G.; Gao, X.; Yang, K.; Li, D.; Kränkel, C. Multilayer black phosphorus as saturable absorber for an Er:Lu_2O_3 laser at ~3 μm. *Photon. Res.* **2016**, *4*, 181–186. [CrossRef]
31. Feng, C.; Ma, B.; Qiao, W.; Li, G.; Zhao, J.; Yang, K.; Li, D.; Li, G.; Zhao, S.; Li, T. Passively Q-switched Er:Lu_2O_3 laser at 2.8 μm with TiC saturable absorber. *Appl. Optics* **2020**, *59*, 8066–8070. [CrossRef]
32. Yan, Z.; Li, T.; Zhao, J.; Yang, K.; Li, D.; Li, G.; Fan, M.; Zhang, S.; Zhao, S. Passively Q-switched 2.85 μm Er:Lu_2O_3 laser with WSe_2. *Laser Phys. Lett.* **2018**, *15*, 085802. [CrossRef]
33. Feng, C.; Qiao, W.; Liu, Y.; Huang, J.; Liang, Y.; Zhao, Y.; Song, Y.; Li, T. Modulation of MXene Nb_2CT_x saturable absorber for passively Q-switched 2.85 μm Er:Lu_2O_3 laser. *Opt. Lett.* **2021**, *46*, 1385–1388. [CrossRef] [PubMed]
34. Li, G.; Li, T.; Qiao, W.; Feng, T.; Feng, C.; Zhao, J.; Li, G.; Zhao, S. Passively Q-switched Er:Lu_2O_3 laser with MXene material $Ti_4N_3T_x$ (T = F, O, or OH) as a saturable absorber. *Opt. Lett.* **2020**, *45*, 4256–4259. [CrossRef] [PubMed]
35. Fan, M.; Li, T.; Zhao, S.; Li, G.; Ma, H.; Gao, X.; Kränkel, C.; Huber, G. Watt-level passively Q-switched Er:Lu_2O_3 laser at 2.84 μm using MoS_2. *Opt. Lett.* **2016**, *41*, 540–543. [CrossRef] [PubMed]
36. Yao, Y.; Cui, N.; Wang, Q.; Dong, L.; Liu, S.; Sun, D.; Zhang, H.; Li, D.; Zhang, B.; He, J. Highly efficient continuous-wave and $ReSe_2$ Q-switched ~3 μm dual-wavelength Er:YAP crystal lasers. *Opt. Lett.* **2019**, *44*, 2839–2842. [CrossRef]
37. Fan, M.; Li, T.; Zhao, J.; Zhao, S.; Li, G.; Yang, K.; Su, L.; Ma, H.; Kränkel, C. Continuous wave and ReS_2 passively Q-switched Er:SrF_2 laser at ~3 μm. *Opt. Lett.* **2018**, *43*, 1726–1729. [CrossRef]
38. Liu, J.; Huang, H.; Zhang, F.; Zhang, Z.; Liu, J.; Zhang, H.; Su, L. Bismuth nanosheets as a Q-switcher for a mid-infrared erbium-doped SrF_2 laser. *Photonics Res.* **2018**, *6*, 762–767. [CrossRef]

nanomaterials

MDPI

Article

Automated Prediction of Bacterial Exclusion Areas on SEM Images of Graphene–Polymer Composites

Shadi Rahimi [1,†], Teo Lovmar [1,†], Alexandra Aulova [2], Santosh Pandit [1], Martin Lovmar [1,3], Sven Forsberg [4], Magnus Svensson [3], Roland Kádár [2] and Ivan Mijakovic [1,5,*]

1 Division of Systems and Synthetic Biology, Department of Life Sciences, Chalmers University of Technology, 41296 Gothenburg, Sweden; shadir@chalmers.se (S.R.)
2 Division of Engineering Materials, Chalmers University of Technology, 41296 Gothenburg, Sweden
3 Wellspect Healthcare, Aminogatan 1, 43121 Mölndal, Sweden
4 2D Fab AB, Bultgatan 20, 85350 Sundsvall, Sweden
5 The Novo Nordisk Foundation Center for Biosustainability, Technical University of Denmark, 2800 Kongens Lyngby, Denmark
* Correspondence: ivan.mijakovic@chalmers.se
† These authors contributed equally to this work.

Abstract: To counter the rising threat of bacterial infections in the post-antibiotic age, intensive efforts are invested in engineering new materials with antibacterial properties. The key bottleneck in this initiative is the speed of evaluation of the antibacterial potential of new materials. To overcome this, we developed an automated pipeline for the prediction of antibacterial potential based on scanning electron microscopy images of engineered surfaces. We developed polymer composites containing graphite-oriented nanoplatelets (GNPs). The key property that the algorithm needs to consider is the density of sharp exposed edges of GNPs that kill bacteria on contact. The surface area of these sharp exposed edges of GNPs, accessible to bacteria, needs to be inferior to the diameter of a typical bacterial cell. To test this assumption, we prepared several composites with variable distribution of exposed edges of GNP. For each of them, the percentage of bacterial exclusion area was predicted by our algorithm and validated experimentally by measuring the loss of viability of the opportunistic pathogen *Staphylococcus epidermidis*. We observed a remarkable linear correlation between predicted bacterial exclusion area and measured loss of viability (R^2 = 0.95). The algorithm parameters we used are not generally applicable to any antibacterial surface. For each surface, key mechanistic parameters must be defined for successful prediction.

Keywords: antibacterial; bacterial exclusion area; graphene flakes; algorithm; vertical

1. Introduction

Misuse of antibiotics and the global rise of antibiotic resistance is heralding a post-antibiotic era in which humanity will need to find alternatives to classical antibiotics. One venue that is being extensively explored is the discovery of new materials resulting in surface structures with antibacterial properties. These comprise natural and artificial fabricated surfaces. Natural biocidal surfaces include insect wings with nanopillars (cicada wing [1–3]) and animal skins with nanostructures (shark and gecko skins [4,5]). In addition to these, a variety of chemical and mechanical methods were used to replicate these naturally occurring surfaces [6–9]. Nanostructures on metals such as Cu, Al, Ti, and Au [10–15], TiO_2 [16], silica and Al_2O_3 [17], black silicon [18], polymers [19], and nanocomposites such as Ti/Au [20] and Ag/polylactide [21] are examples of these fabricated surfaces. All these nanostructures have various mechanisms of antibacterial action but summarily result in reduced attachment of bacterial biofilms. The key limitation in this process is our ability to measure the antibacterial efficiency of newly developed surfaces rapidly and reliably. Proper microbiology assays which involve assessment of cell viability are needed for this, and these typically require days to weeks. SEM imaging is a valuable tool to assess surfaces modified by nanostructures. However, the information that we can provide by SEM observation is only qualitative information on the density and positioning of the flakes.

To address this bottleneck, we attempted to develop an automated pipeline for the prediction of antibacterial potential of a surface based on scanning electron microscopy (SEM) images. We propose that in order to be efficient, such algorithms cannot be overly generic and must take into account a certain level of mechanistic understanding of the antibacterial properties of the specific surface. To test this assumption, we developed an example involving a polymer composite containing oriented exfoliated graphite nanoflakes, recently reported by our group [7]. The orientation and density of distribution of exfoliated graphite nanoplatelets within our polymer nanocomposite can be controlled in such a way as to reduce bacterial viability by a factor of 99.9999%, which is currently the benchmark in the field [7]. The key antibacterial property of this material is the density of sharp exposed edges of exfoliated graphite flakes that are accessible on the surface and kill bacteria on contact. The mechanism of interaction of graphene-based materials, including exfoliated graphene, with bacterial cells includes mechanical damage, electron transfer, insertion, lipid extraction, pore formation, and wrapping of cells [22]. Graphene material insertion, lipid extraction, and pore formation are classified as mechanical damage to the cell. Lipid peroxidation and electron transfer cause oxidative stress. Masking mode could also be the underlying basis of the wrapping mechanism.

In general, larger size, sharper edge, and aggregation are advantageous to inserting/cutting, lipid extraction, and pore formation modes, resulting in stronger destabilization of membrane [23]. The size of graphene nanosheets positively correlated with the extent of lipid extraction. Larger graphene flakes had more phospholipid extraction power, thereby demonstrating stronger antibacterial activity compared to smaller flakes [24]. Small-sized graphene materials induce oxidative stress because the density of defects increases with reduction of the size of the materials [23]. The wrapping or masking mechanism also requires large lateral size, specifically, of micrometer-sized graphene materials [25], as smaller graphene could easily pierce the phospholipid membrane [26]. Wrapping can basically be ruled out with nanoflakes embedded in a polymer matrix. The orientation of graphene materials could also be engineered to confer superior antibacterial property [27]. Magnetic field and chemical vapor deposition methods are used to control the orientation of graphene materials [28,29]. Our group demonstrated that vertically aligned graphene materials are more lethal for bacteria than randomly orientated ones [7]. This might be due to the synergistic effect of physical puncturing of cell membrane and effective electron transfer, as vertical graphene and cell membrane are in contact with each other efficiently [30].

In a nutshell, for a particular antibacterial surface to work, the surface area of sharp exposed edges of exfoliated graphite flakes accessible to bacteria needs to be inferior to the diameter of a typical bacterial cell (Figure 1).

To develop and test our predictor of the antimicrobial effectiveness of these polymer nanocomposites, we devised an algorithm that analyzes SEM images of the surface, identifies exposed (accessible) edges of exfoliated graphene, predicts bacterial exclusion areas around the edges (based on bacterial diameter), and calculates the remaining area available to bacterial cells (Figure 1). Next, we prepared six composite samples in which the distribution and orientation of exfoliated graphite nanoflakes varied considerably, and we measured the loss of bacterial viability on each surface. SEM images of those same surfaces were randomly obtained and analyzed by our algorithm. The correlation between predicted bacterial exclusion area and the measured loss of viability for each surface turned out to be proportional, with an R^2 coefficient of 0.95. An overview of the analysis presented in this study is shown in Figure 1. We concluded that prediction of antibacterial potential from SEM images is possible for this type of material. We argue that our approach is generally applicable; however, for other types of surfaces, the approach will be valid only if the antibacterial mechanism is understood well enough to include the key parameters in the prediction.

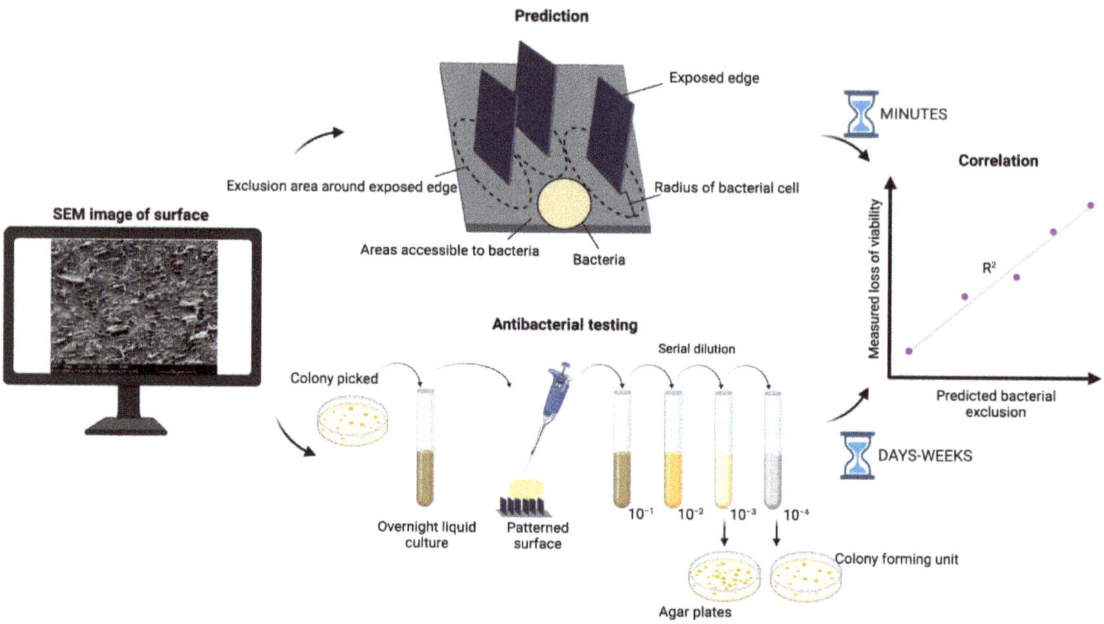

Figure 1. Overview of our comparison of prediction with experimental results. SEM image analysis of the materials is the input. Antibacterial testing, based on counting of colony-forming units (CFUs), can require days to weeks. Prediction based on SEM image analysis and assessment of bacterial exclusion area is carried out in minutes. Predicted and measured antibacterial effect for each tested surface is compared and correlated. Figure was created with BioRender.com.

2. Results and Discussion

2.1. Algorithm Predicts Bacterial Exclusion Areas from SEM Images of Composite Surface

We fabricated 6 polymer composite samples with varying density of graphene flakes (5, 10, 15, 20%) sliced in a longitudinal (L) and/or transversal (T) direction (Table 1). For all samples, SEM images were randomly obtained and submitted to computational analysis. To assess the antibacterial property of the samples, software was developed to implement an algorithm for predicting and quantifying the bacteria exclusion area, defined as a fixed radius around each detected graphene edge (Figure 2). The algorithm identifies flake edges (Figure 2a), identifies zones in which flake edges are vertically oriented and exposed enough to have antibacterial effects (Figure 2b), defines the exclusion zone as a perimeter around each edge that is not accessible to the bacteria (Figure 2c), and, finally, predicts the bacterial exclusion area as a % of the surface not accessible to bacteria (Figure 2d). It should be noted that the graphene-integrated area is not necessarily the same as the bacterial exclusion area of the graphene-integrated polymer surface. In fact, the bacterial exclusion area is the area where the attachment of bacteria is impossible due to the exposed edges of exfoliated graphene, and not due simply to the presence of graphene within the polymer. The boundary of bacterial exclusion area is determined based on geometrical assumptions that the bacteria are not able to attach in the vicinity of these exposed sharp graphene edges. Thus, the radius of the nonattachment area is determined based on the average size of bacteria, which is 0.9 µm in the case of *Staphylococcus epidermidis*. The basic concept behind the bacterial exclusion area in the algorithm is that the area with exposed sharp graphene edges is a "killing zone" for bacteria since the graphene edges disrupt the membrane of bacterial cells and thereby confer an antibacterial property to the material. The concept might be generalized to other materials if they exhibit the same mechanism of physical damage to bacterial cells as graphene.

Table 1. Comparing loss of viability of *S. epidermidis* from entire surface with varying density of graphene flakes (5, 10, 15, 20%) sliced in longitudinal (L) and/or transversal (T) direction. The average of predicted bacterial exclusions was calculated from several images characterized by graphene flakes randomly obtained at different magnifications. The generated outputs were measured using these parameters: minimum_edge_area = 5 and exclusion_radius = 0.9. SEM images were obtained from random positions on 8 different samples. The first four images from two samples have no proper orientation. The loss of viability of *S. epidermidis* measured from whole surface of sample was obtained from our previous study [7]. The data represent the mean ± SD of independent replicates with repetitions.

Image	Contrast Threshold for a Valid Edge	Average of Predicted Bacterial Exclusion from Images from Different Positions in One Sample (%)	Magnification	Measured Loss of Viability (%)
M25-6.tif	50	9.73 ± 1.45	1000×	−16 ± 42.4
M25-2.tif	50		1000×	
23.tif	80	20.61 ± 5.62	1000×	13.518 ± 19.49
11.tif	90		1400×	
5%-L2.tif	40	35.97 ± 3.7	1000×	28.18 ± 7.3
5%-L-3.tif	40		500×	
5%-L-5.tif	40		650×	
5%-L1.tif	40		1000×	
5%-L4.tif	20		1000×	
5%-T1.tif	25	30.58 ± 0.58	2500×	40.80 ± 10.48
5%-T2.tif	15		5000×	
5%-T4.tif	20		5000×	
10%-L1.tif	30	61.85 ± 4.46	800×	61.12 ± 4.9
10%-L2.tif	28		800×	
10%-L3.tif	20		1500×	
10%-L4.tif	20		1500×	
15%-T1.tif	25	73.85 ± 6.84	1500×	78.54 ± 7.98
15%-T2.tif	10		5000×	
15%-T3.tif	25		1000×	
15%-T4.tif	7		5000×	
15%-T6.tif	12		2500×	
20%-T1.tif	15	84.23 ± 5.33	2500×	94.51 ± 0.66
20%-T2.tif	8		5000×	
20%-T3.tif	23		1000×	
20%-T4.tif	10		2500×	
20%-T5.tif	8		5000×	
15%-L1.tif	30	89.86 ± 2.91	1000×	99.99 ± 0
15%-L2.tif	30		1000×	
15%-L3.tif	22		1200×	
15%-L4.tif	20		1200×	
15%-L5.tif	25		800×	

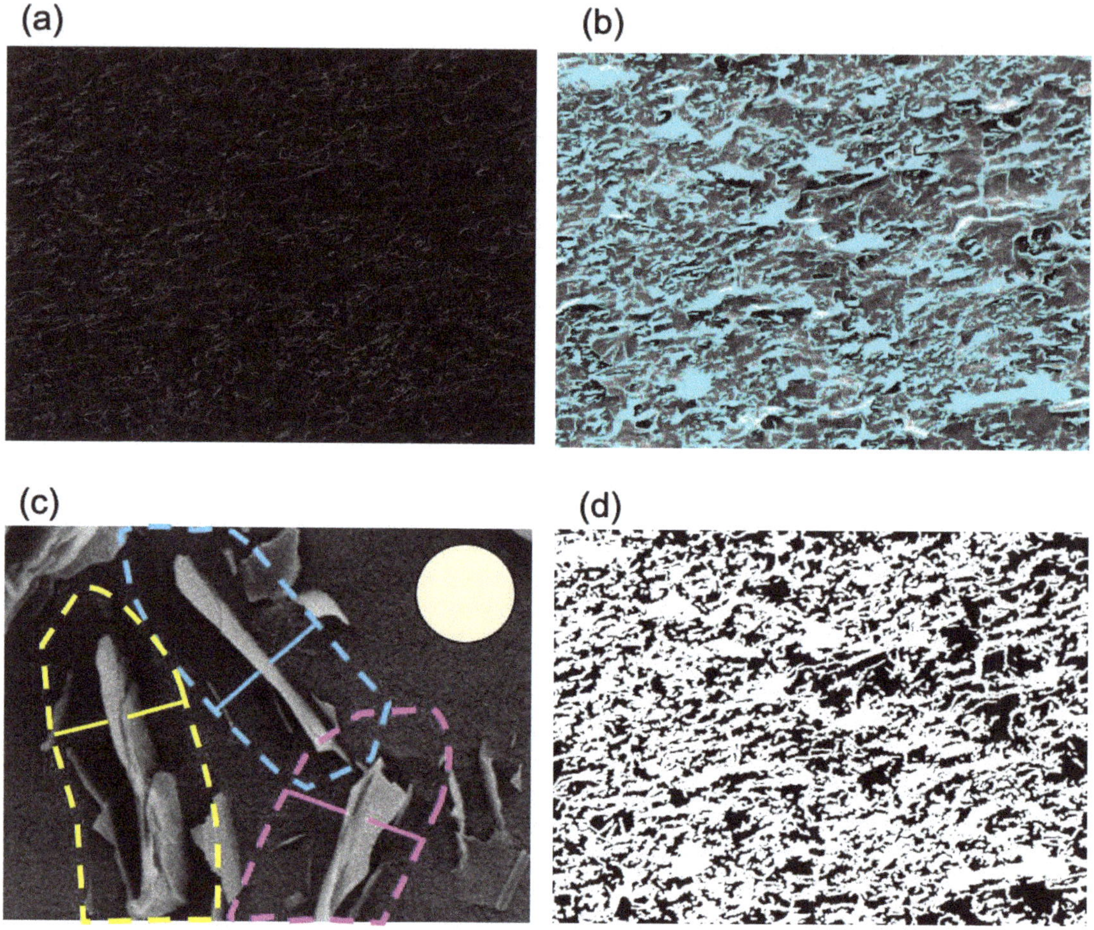

Figure 2. Automated prediction of bacterial exclusion areas. (**a**) Detection of graphene flake edges, (**b**) definition of areas with correct graphene orientation (marked in blue), (**c**) the exclusion zone around each edge (dotted line with exclusion radius = 0.9 µm) that is not accessible to the bacteria (tan disc shows a schematic *S. epidermidis* cell with 1.5 µm diameter), and (**d**) predicted bacterial exclusion areas (marked in white) on the analyzed surface.

The software is available from this link: https://github.com/SysBioChalmers/bacterial-exclusion-prediction (created on 21 February 2023). There are different versions of the software for different operating systems, such as Linux, MacOS, or Windows, that can be installed as explained in the installation section. The input for the algorithm is an image in "tif" format. The software makes use of Tesseract, a robust text recognition algorithm [31], and can automatically read the text showing the magnification written on SEM images. Thus, the software can adapt the results to the scale and the estimated size, provided in µm, of the flakes.

Several parameters can be controlled in the "Bacteria exclusion" section of the web interface. One can adjust the contrast threshold and the minimum area (pixel^2) for identifying an edge. The radius around the graphene edge (μm) to exclude bacteria can also be adjusted to accommodate bacterial cells of different sizes or shapes [23]. We recommend using images with magnification between 1000×–2500×. The algorithm can be run in batch mode, allowing the user to specify a directory from which all existing images are analyzed together, and the output is stored inside an automatically generated subdirectory, "output", so as not to clutter the working directory.

Flake sharpness, properly oriented flakes, and predicted areas contributing to bacterial exclusion from the surface could be visualized in the algorithm output files, including "edge_sharpness.png", "graphene.png", and "bacteria-exclusion.png" files, respectively (Figure 2a, Figure 2b, and Figure 2d). The "edge_sharpness.png" is derived by calculating the sum of the absolute contrast for each pixel in each of the four directions (as explained in the section covering the bacterial exclusion area in Materials and Methods). This provides an image of the sharpness at each pixel. The "graphene.png" image is derived from the contrast by using a threshold and then identifying connected regions. These regions, or flakes, are then filtered by minimum area to remove smaller noise. Finally, the "bacteria-exclusion.png" image is created by calculating if the distance to the closest flake is below a threshold or not. The algorithm also visualizes the percentage of excluded bacteria area (the percentage of white pixels in Figure 2d) in relation to the terminal.

2.2. Predicted Bacterial Exclusion Areas Correlate Well with Experimental Measurements of *S. epidermidis* Viability

The percentage of bacterial exclusion areas predicted by the algorithm was compared to measured antibacterial effects for all tested surfaces [7]. While the CFU counting method provides an aggregate measure for the entire tested surface (typically a sample area of 0.5 cm^2), for practical reasons, the algorithm prediction was obtained from a sample of SEM images that are representative, but do not cover the entire sample surface (5 SEM images from random positions, covering a total area of 0.1–0.2 mm^2). The average % of bacterial exclusion from these SEM images was compared with the measured aggregate % of loss of viability. As shown in Figures 3–5, to obtain the proper bacterial exclusion value, the user is required to adjust the "contrast threshold for a valid edge" to cover all the flakes, as indicated in Table 1. It is recommended to choose a lower contrast threshold, in the range of 7–25, for images with higher magnification (2500×–5000×) (Table 1). However, a higher contrast threshold, of 20–40, would be required for images obtained at lower magnification (500×–1500×) (Table 1).

As we previously investigated [7], the ratio of graphite nanoplatelets to total surface was significantly different among longitudinally (L) and transversally (T) sliced samples with ≥15% integrated graphite nanoplatelets. The ratio of graphite nanoplatelets to total surface was significantly higher in longitudinally (L) sliced samples compared to transversally (T) sliced samples with the same density of graphite nanoplatelets used for fabrication. Subsequently, there was an clear difference in loss of *Staphylococcus epidermidis* viability among these samples, as longitudinally (L) sliced polymers with ≥15% graphite nanoplatelets showed higher loss of viability compared to the transversally (T) sliced samples. Thus, the increased ratio of graphite nanoplatelets to total surface in longitudinally (L) sliced polymers could result in increased loss of viability. Interestingly, these results were clearly reflected by our prediction and, as it was shown (Table 1), the average predicted bacterial exclusion of 15%-L images was 89.86% ± 2.91%, which was significantly higher than that of 15%-T images (73.85% ± 6.84%).

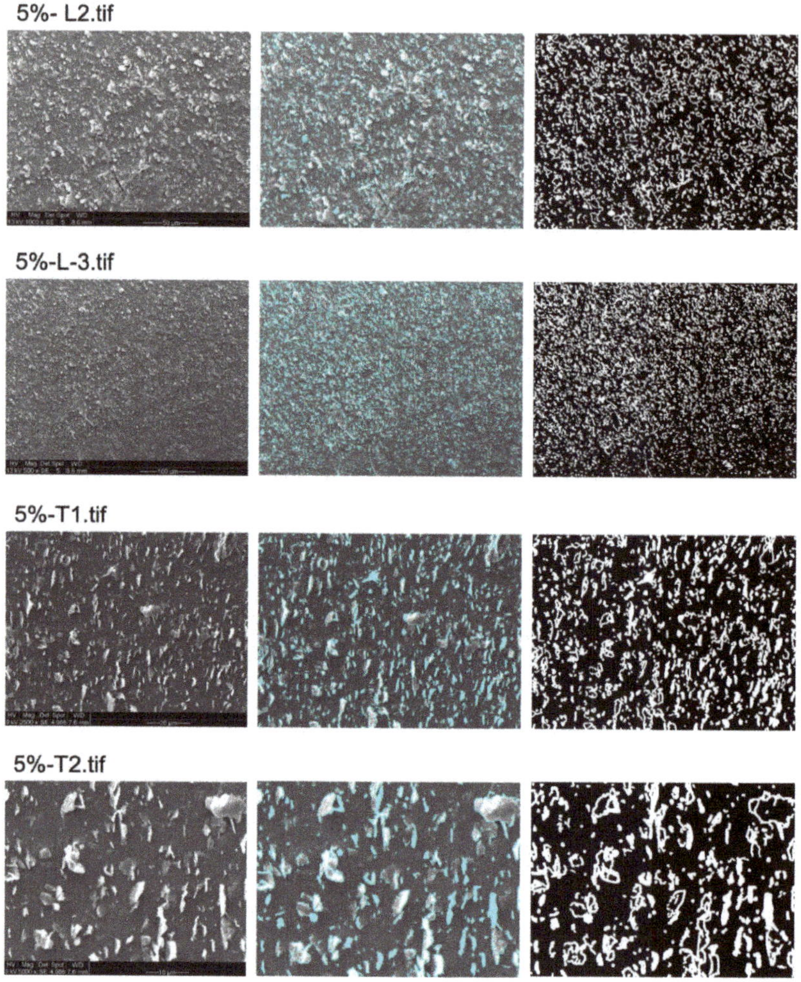

Figure 3. Prediction of areas contributing to bacterial exclusion from 2 different images at different magnifications from 5%-L and 5%-T samples. Proper graphene-oriented areas on original image (**left**) marked in cyan (**middle**) and predicted bacterial exclusion areas marked in white (**right**) on the surface of samples. The original images were adapted with permission from Ref. [7]. 2023, Shadi Rahimi.

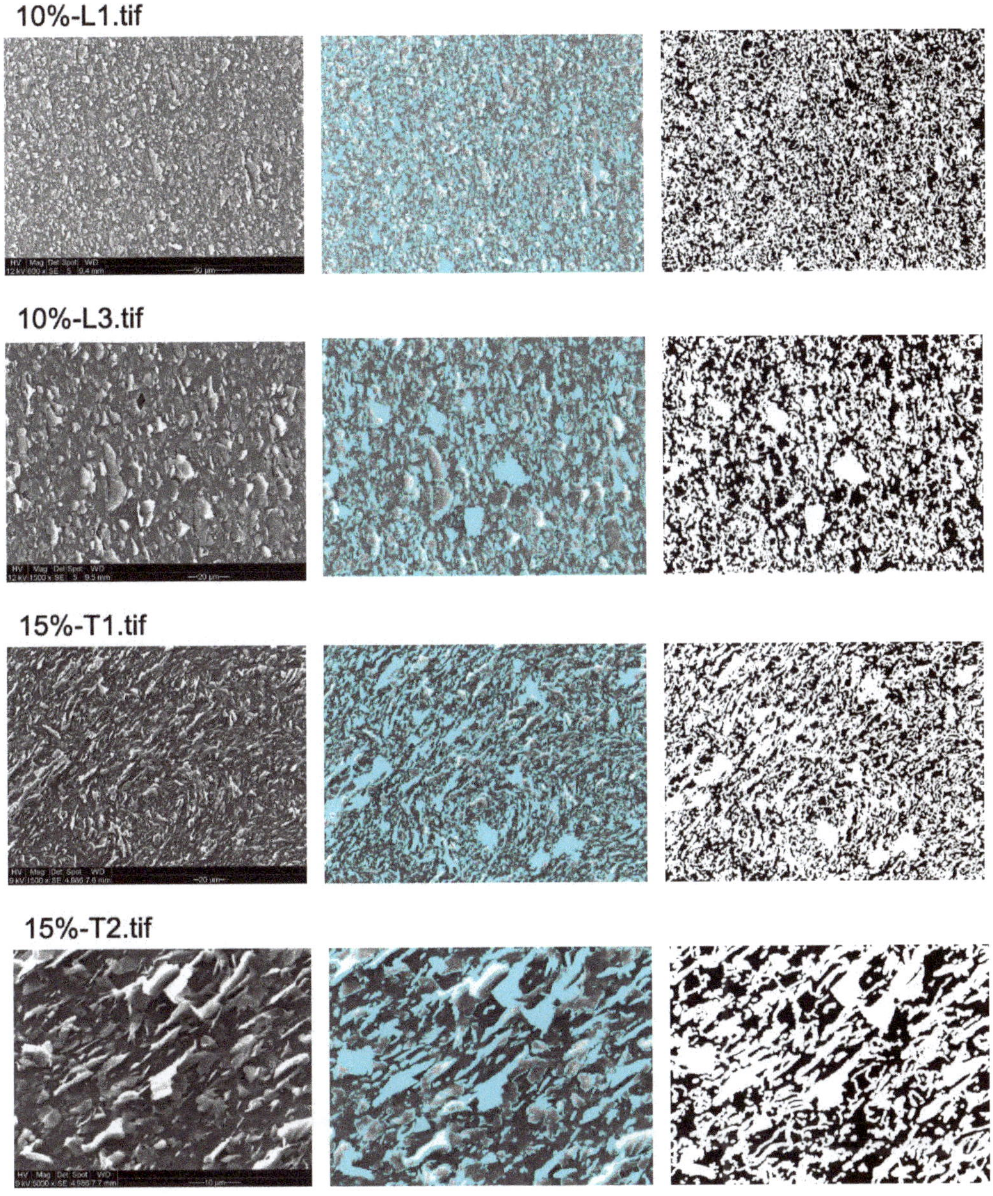

Figure 4. Prediction of areas contributing to bacterial exclusion from 2 different samples of 10%-L and 15%-T. Two images at random locations and at different magnifications were obtained from each sample. Original image is shown on left, proper graphene-oriented areas on the surface of samples are marked in blue in middle image, and predicted bacterial exclusion areas are marked in white on the right side. The original images were adapted with permission from Ref. [7]. 2023, Shadi Rahimi.

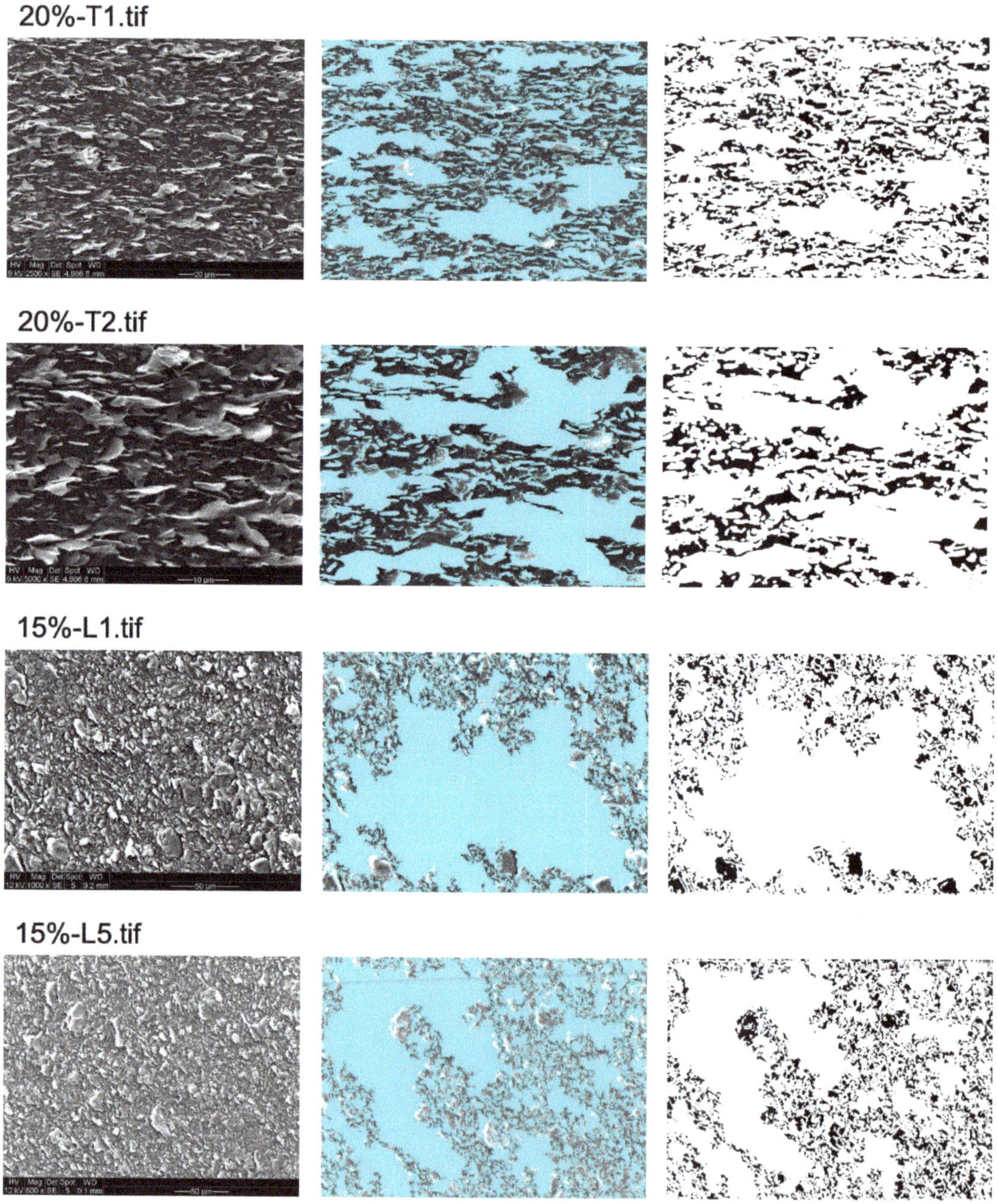

Figure 5. Prediction of areas contributing to bacterial exclusion from 2 different samples of 20%-T and 15%-L. Two images at random locations and at different magnifications were obtained from each sample. Original image is shown on left, proper graphene-oriented areas on the surface of samples are marked in blue in middle image, and predicted bacterial exclusion areas are marked in white on the right side. The original images were adapted with permission from Ref. [7]. 2023, Shadi Rahimi.

As indicated in Table 1, the average % of bacterial exclusion area predicted by the algorithm matches the trend of the measured loss of viability of *S. epidermidis* cells in all six samples. In addition to the six samples, we have included the first two samples in Table 1. The surfaces of the latter have no proper orientation and show low loss of viability. High heterogeneity on the surface of these samples results in high standard deviation in measured loss of viability; thus, we excluded them from the correlation plot. Based on the results from Table 1, a linear correlation plot was drawn for the predicted bacterial exclusion versus measured loss of viability of *S. epidermidis* (Figure 6). The predicted bacterial exclusion and the measured loss of viability of *S. epidermidis* correlate extremely well, with an R-squared value (R^2) of 0.95. Our conclusion was that the algorithm can be used to provide reliable estimates of antibacterial protection offered by this specific type of composite surface.

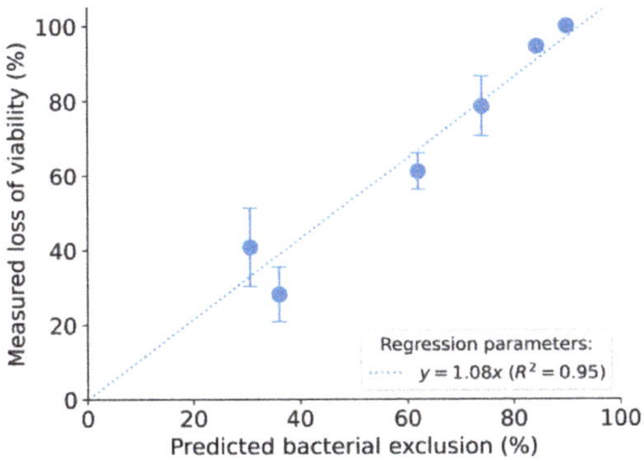

Figure 6. The linear correlation plot for the predicted bacterial exclusion versus measured loss of viability of *S. epidermidis*.

3. Conclusions

In this study, we propose that the key parameter for the antibacterial activity of this type of material is the density of distribution of sharp exposed edges of exfoliated graphite flakes that are accessible on the surface and can kill bacteria on contact. Based on this assumption and using SEM images of the surface, we designed an in silico predictor of the antibacterial potential of various exfoliated graphite nanocomposites. The predictor demonstrated a very strong correlation between the actual antibacterial potential of different surfaces (measured experimentally) and the density and orientation of exposed nanoflakes. We propose that the presented approach has the potential to drastically speed up the investigation of new materials with integrations with the potential to prevent bacterial attachment. Thus, it is possible to devise an algorithm that can automatically detect key surface properties from SEM images and, based on this, accurately predict the potential of the surface to offer antibacterial protection. Such in silico predictions based on SEM images can be used to screen a large number of materials before selecting the most promising ones for antibacterial testing (which typically lasts from days to weeks). While we would like to argue that this approach can be of general utility in the field of antibacterial surfaces, we wish to stress that its proper use critically depends on two factors. Firstly, the antibacterial mechanism of the given surface must be sufficiently understood to program the algorithm to recognize the key surface properties. In the presented case, these key features were the orientation and distribution of exposed graphene edges. Secondly, the uniformity of the surface features across the entire sample may play a key role. For materials with less uniform features, it would be advisable to cover larger areas with SEM analysis, which may constitute a tradeoff regarding time (acquisition of SEM images) and computational resources. Within these limits, we would argue that the presented approach has the potential to speed up the investigation of new materials with integrations with the potential to prevent bacterial attachment and curb bacterial infections [4].

4. Materials and Methods

4.1. Polymer Integration with Graphene Nanoplatelets and Scanning Electron Microscopy (SEM) Imaging

The low-density polyethylene (LDPE) graphene nanoplatelet nanocomposites were prepared as described previously [7]. LDPE was utilized as the composite matrix. M25 graphene nanoplatelets from 2D Fab (Sweden) with average particle diameter of 25 microns and thickness of 6–8 nm were used as a filler. LDPE pellets were cryogenically ground into powder form and mixed with well-dispersed and homogenized suspension of graphene nanoplatelets with acetone. This process was followed by drying at 60 °C. Then, the extrusion process of nanocomposites was carried out using a circular die by Brabender 19/25 D single-screw extruder (Duisburg, Germany) by means of a compression screw (diameter D = 19 mm and screw length of 25 × 19, compression ratio 2:1). Samples for antibacterial analysis and SEM imaging were collected. SEM imaging was performed randomly from the entire surface using Supra 60 VP microscope (Carl Zeiss AG, Oberkochen, Germany).

4.2. Evaluation of Antibiofilm Potential

The antibacterial activity of graphene-integrated materials was tested against the opportunistic bacterial pathogen *S. epidermidis* as a model for Gram-positive bacterium [7]. The overnight culture of *S. epidermidis* bacteria was diluted in fresh tryptic soy broth (TSB) (Sigma Aldrich, Stockholm, Sweden) to obtain the final inoculum of 2–5 × 10^6 CFU/mL and seeded in the pre-sterilized integrated and nonintegrated surfaces. Samples with bacterial inoculum were incubated at 37 °C for 6 h without agitation for the formation of biofilms. After 24 h of growth, samples were collected in 5 mL of 0.89% of sodium chloride solution for viability test. The biofilms were detached from the surface and homogenized by sonication probe at 10% for 30 s. The homogenized biofilm suspensions were serially diluted into 0.89% of sodium chloride solution and plated onto the LB agar plates. Agar plates were incubated at 37 °C for 24 h and the number of colonies was counted. The number of colonies grown on integrated surfaces divided by the number of colonies grown on control nonintegrated surfaces multiplied by 100 were expressed as the percentage of viability; 100 minus the percentage of viability equals percentage of loss of viability.

4.3. Development of SEM Image Analysis Algorithm

A software implementing an algorithm for analyzing SEM images from surfaces with graphene orientation was developed. The algorithm was written in the Rust programming language (rust-lang.org) (accessed on 20 June 2022). The algorithm is packed with a graphical user interface for simple parameter testing and a batch analysis feature for efficient screening through a web interface which can be reached through http://127.0.0.1:8080 once the algorithm is up and running. The software is available from this link: https://github.com/SysBioChalmers/bacterial-exclusion-prediction (created on 21 February 2023).

4.4. Bacterial Exclusion Area Determination

The source pictures for the analysis were obtained by means of SEM at the fixed settings. Due to different surface topology and conductivity (due to different conductive filler dispersion), the pictures had different values of contrast and brightness. All source pictures were provided in 8-bit pixel form (values ranging from 0 to 255).

The bacterial exclusion area is the area where the attachment of bacterial film is impossible. It is determined, based on geometrical assumptions, that the bacteria will not be able to attach in the vicinity of the sharp graphene edge. The radius of nonattachment is determined by the average size of bacteria; for example, *S. epidermidis* size was determined as 0.9 µm.

The algorithm operates as following:

1. Graphene edge detection. This procedure searches for the "sharp" contrast area in the image. Pixel intensities, I, of surrounding pixels were obtained and the absolute intensity difference of pixels positioned opposite from each other with respect to the analyzed pixel was averaged as shown in Figure 7. Value ranging from 0 to 255 was obtained for each pixel, which represents the "absolute" gradient of the intensity of the image and is denoted as C further on:

$$C = \frac{1}{4}(|I_N - I_S| + |I_W - I_E| + |I_{NW} - I_{SE}| + |I_{NE} - I_{SW}|) \quad (1)$$

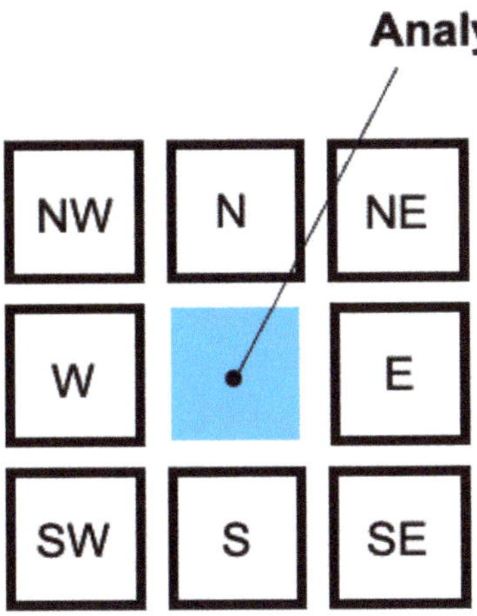

Figure 7. Schematic representation of intensity calculation for edge detection.

2. Thresholding is carried out by filtering pixels with a preset value of intensity C_0. This value can be set by user. Binary image is the output of this operation.

3. Denoising is performed according to the following subroutine:

a. The binary image enables us to define groups of connected white pixels isolated from other pixels by black pixels as individual groups. Each group represents a detected graphene flake.
b. Calculation of the area of each group by counting the pixels in the group.
c. Remove group if the area is below a threshold set by the user.

4. Calculation of the Euclidean distance for every pixel of the image to the closest white pixel. Next, the threshold determined by the size of the bacteria is applied to the calculated value. This operation results in a binary image.

Author Contributions: Conceptualization, I.M.; methodology, T.L.; validation, T.L. and M.L.; investigation, S.R.; resources, S.F.; writing—original draft preparation, S.R.; writing—review and editing, T.L., A.A., S.P., M.L., R.K., and I.M.; supervision, I.M.; funding acquisition, M.L and M.S. All authors have read and agreed to the published version of the manuscript.

Funding: This work was supported by a grant from Vinnova—Sveriges innovationsmyndighet [2020-00792], the Novo Nordisk Foundation [NNF20OC0064547], Kristina Stenborgs foundation for scientific research [C 2021-1705], and Vetenskapsrådet [2020-04096].

Data Availability Statement: The software is available from this link: https://github.com/SysBioChalmers/bacterial-exclusion-prediction (created on 21 February 2023).

Conflicts of Interest: The authors declare no conflict of interest.

References

1. Román-Kustas, J.; Hoffman, J.B.; Reed, J.H.; Gonsalves, A.E.; Oh, J.; Li, L.; Hong, S.; Jo, K.D.; Dana, C.E.; Miljkovic, N.; et al. Molecular and topographical organization: Influence on cicada wing wettability and bactericidal properties. *Adv. Mater. Interfaces* **2020**, *7*, 2000112. [CrossRef]
2. Ivanova, E.P.; Hasan, J.; Webb, H.K.; Truong, V.K.; Watson, G.S.; Watson, J.A.; Baulin, V.A.; Pogodin, S.; Wang, J.Y.; Tobin, M.J.; et al. Natural bactericidal surfaces: Mechanical rupture of Pseudomonas aeruginosa cells by cicada wings. *Small* **2012**, *8*, 2489. [CrossRef] [PubMed]
3. Bandara, C.D.; Singh, S.; Afara, I.O.; Wolff, A.; Tesfamichael, T.; Ostrikov, K.; Oloyede, A. Bactericidal Effects of Natural Nanotopography of Dragonfly Wing on *Escherichia coli*. *ACS Appl. Mater. Interfaces* **2017**, *9*, 6746. [CrossRef] [PubMed]
4. Rostami, S.; Garipcan, B. Evolution of antibacterial and antibiofouling properties of sharkskin-patterned surfaces. *Surf. Innov.* **2022**, *10*, 165. [CrossRef]
5. Watson, G.S.; Green, D.W.; Schwarzkopf, L.; Li, X.; Cribb, B.W.; Myhra, S.; Watson, J.A. A gecko skin micro/nano structure–A low adhesion, superhydrophobic, anti-wetting, self-cleaning, biocompatible, antibacterial surface. *Acta Biomater.* **2015**, *21*, 109. [CrossRef]
6. Pandit, S.; Cao, Z.; Mokkapati, V.R.S.S.; Celauro, E.; Yurgens, A.; Lovmar, M.; Westerlund, F.; Sun, J.; Mijakovic, I. Vertically aligned graphene coating is bactericidal and prevents the formation of bacterial biofilms. *Adv. Mater. Interfaces* **2018**, *5*, 1701331. [CrossRef]
7. Pandit, S.; Gaska, K.; Mokkapati, V.R.S.S.; Celauro, E.; Derouiche, A.; Forsberg, S.; Svensson, M.; Kádár, R.; Mijakovic, I. Precontrolled alignment of graphite nanoplatelets in polymeric composites prevents bacterial attachment. *Small* **2020**, *16*, 1904756. [CrossRef]
8. Sun, J.; Rattanasawatesun, T.; Tang, P.; Bi, Z.; Pandit, S.; Lam, L.; Wasén, C.; Erlandsson, M.; Bokarewa, M.; Dong, J.; et al. Insights into the Mechanism for Vertical Graphene Growth by Plasma-Enhanced Chemical Vapor Deposition. *ACS Appl. Mater. Interfaces* **2022**, *14*, 7152. [CrossRef]
9. Gaska, K.; Kádár, R.; Xu, X.; Gubanski, S.; Müller, C.; Pandit, S.; Mokkapati, V.R.S.S.; Mijakovic, I.; Rybak, A.; Siwek, A.; et al. Highly structured graphene polyethylene nanocomposites. *AIP Conf. Proc.* **2019**, *2065*, 030061.
10. Reed, J.H.; Gonsalves, A.E.; Román, J.K.; Oh, J.; Cha, H.; Dana, C.E.; Toc, M.; Hong, S.; Hoffman, J.B.; Andrade, J.E.; et al. Ultrascalable multifunctional nanoengineered copper and aluminum for antiadhesion and bactericidal applications. *ACS Appl. Bio Mater.* **2019**, *2*, 2726. [CrossRef]
11. Bhadra, C.M.; Khanh Truong, V.; Pham, V.T.; Al Kobaisi, M.; Seniutinas, G.; Wang, J.Y.; Juodkazis, S.; Crawford, R.J.; Ivanova, E.P. Antibacterial titanium nano-patterned arrays inspired by dragonfly wings. *Sci. Rep.* **2015**, *5*, 16817. [CrossRef] [PubMed]
12. Lorenzetti, M.; Dogsa, I.; Stosicki, T.; Stopar, D.; Kalin, M.; Kobe, S.; Novak, S. The influence of surface modification on bacterial adhesion to titanium-based substrates. *ACS Appl. Mater. Interfaces* **2015**, *7*, 1644. [CrossRef] [PubMed]
13. Tsimbouri, P.M.; Fisher, L.; Holloway, N.; Sjostrom, T.; Nobbs, A.H.; Meek, R.D.; Su, B.; Dalby, M.J. Osteogenic and bactericidal surfaces from hydrothermal titania nanowires on titanium substrates. *Sci. Rep.* **2016**, *6*, 36857. [CrossRef]
14. Sengstock, C.; Lopian, M.; Motemani, Y.; Borgmann, A.; Khare, C.; Buenconsejo, P.J.S.; Schildhauer, T.A.; Ludwig, A.; Köller, M. Structure-related antibacterial activity of a titanium nanostructured surface fabricated by glancing angle sputter deposition. *Nanotechnology* **2014**, *25*, 195101. [CrossRef]
15. Wu, S.; Zuber, F.; Brugger, J.; Maniura-Weber, K.; Ren, Q. Antibacterial Au nanostructured surfaces. *Nanoscale* **2016**, *8*, 2620. [CrossRef]
16. Diu, T.; Faruqui, N.; Sjöström, T.; Lamarre, B.; Jenkinson, H.F.; Su, B.; Ryadnov, M.G. Cicada-inspired cell-instructive nanopatterned arrays. *Sci. Rep.* **2014**, *4*, 7122. [CrossRef]
17. Hsu, L.C.; Fang, J.; Borca-Tasciuc, D.A.; Worobo, R.W.; Moraru, C.I. Effect of micro-and nanoscale topography on the adhesion of bacterial cells to solid surfaces. *Appl. Environ. Microbiol.* **2013**, *79*, 2703. [CrossRef] [PubMed]
18. Ivanova, E.P.; Hasan, J.; Webb, H.K.; Gervinskas, G.; Juodkazis, S.; Truong, V.K.; Wu, A.H.F.; Lamb, R.N.; Baulin, V.A.; Watson, G.S.; et al. Bactericidal activity of black silicon. *Nat. Commun.* **2013**, *4*, 2838. [CrossRef]
19. Dickson, M.N.; Liang, E.I.; Rodriguez, L.A.; Vollereaux, N.; Yee, A.F. Nanopatterned polymer surfaces with bactericidal properties. *Biointerphases* **2015**, *10*, 021010. [CrossRef]
20. Slepička, P.; Fajstavr, D.; Krejčová, M.; Rimpelová, S.; Slepičková Kasálková, N.; Kolská, Z.; Švorčík, V. Biopolymer Composites with Ti/Au Nanostructures and Their Antibacterial Properties. *Pharmaceutics* **2021**, *13*, 826. [CrossRef]
21. Bayraktar, I.; Doganay, D.; Coskun, S.; Kaynak, C.; Akca, G.; Unalan, H.E. 3D printed antibacterial silver nanowire/polylactide nanocomposites. *Compos. B Eng.* **2019**, *172*, 671. [CrossRef]
22. Chen, Y.; Pandit, S.; Rahimi, S.; Mijakovic, I. Interactions Between Graphene-Based Materials and Biological Surfaces: A Review of Underlying Molecular Mechanisms. *Adv. Mater. Interfaces* **2021**, *8*, 2101132. [CrossRef]
23. Perreault, F.; De Faria, A.F.; Nejati, S.; Elimelech, M. Antimicrobial properties of graphene oxide nanosheets: Why size matters. *ACS Nano* **2015**, *9*, 7226. [CrossRef] [PubMed]
24. Tu, Y.; Lv, M.; Xiu, P.; Huynh, T.; Zhang, M.; Castelli, M.; Liu, Z.; Huang, Q.; Fan, C.; Fang, H.; et al. Destructive extraction of phospholipids from *Escherichia coli* membranes by graphene nanosheets. *Nat. Nanotechnol.* **2013**, *8*, 594. [CrossRef]
25. Liu, S.; Hu, M.; Zeng, T.H.; Wu, R.; Jiang, R.; Wei, J.; Wang, L.; Kong, J.; Chen, Y. Lateral dimension-dependent antibacterial activity of graphene oxide sheets. *Langmuir* **2012**, *28*, 12364. [CrossRef] [PubMed]

6. Dallavalle, M.; Calvaresi, M.; Bottoni, A.; Melle-Franco, M.; Zerbetto, F. Graphene can wreak havoc with cell membranes. *ACS Appl. Mater. Interfaces* **2015**, *7*, 4406. [CrossRef]
7. Wei, W.; Li, J.; Liu, Z.; Deng, Y.; Chen, D.; Gu, P.; Wang, G.; Fan, X. Distinct antibacterial activity of a vertically aligned graphene coating against Gram-positive and Gram-negative bacteria. *J. Mater. Chem. B* **2020**, *8*, 6069. [CrossRef]
8. Zheng, K.; Li, K.; Chang, T.H.; Xie, J.; Chen, P.Y. Synergistic Antimicrobial Nanomaterials: Synergistic Antimicrobial Capability of Magnetically Oriented Graphene Oxide Conjugated with Gold Nanoclusters (Adv. Funct. Mater. 46/2019). *Adv. Funct. Mater.* **2019**, *29*, 1904603. [CrossRef]
9. Zheng, W.; Zhao, X.; Fu, W. Review of vertical graphene and its applications. *ACS Appl. Mater. Interfaces* **2021**, *13*, 9561. [CrossRef]
0. Lu, X.; Feng, X.; Werber, J.R.; Chu, C.; Zucker, I.; Kim, J.H.; Osuji, C.O.; Elimelech, M. Enhanced antibacterial activity through the controlled alignment of graphene oxide nanosheets. *Proc. Natl. Acad. Sci. USA* **2017**, *114*, E9793–E9801. [CrossRef]
1. Smith, R. An overview of the Tesseract OCR engine. In Proceedings of the Ninth International Conference on Document Analysis and Recognition (ICDAR 2007), Curitiba, Brazil, 23–26 September 2007; IEEE: New York, NY, USA, 2007; Volume 2, pp. 629–633.

nanomaterials

MDPI

Article

Reconfigurable Single-Layer Graphene Radio Frequency Antenna Device Capable of Changing Resonant Frequency

Hyeon Jun Hwang [1], So-Young Kim [1], Sang Kyung Lee [2] and Byoung Hun Lee [1,*]

[1] Center for Semiconductor Technology Convergence, Department of Electrical Engineering, Pohang University of Science and Technology, Cheongam-ro 77, Nam-gu, Pohang 37673, Gyeongbuk, Republic of Korea; hhjune@postech.ac.kr (H.J.H.); una0918@gmail.com (S.-Y.K.)

[2] Alphagraphene Inc. 77, Cheongam-ro, Nam-gu, Pohang 37673, Gyeongbuk, Republic of Korea; alphagraphene.korea@gmail.com

* Correspondence: bhlee1@postech.ac.kr

Abstract: A reconfigurable passive device that can manipulate its resonant frequency by controlling its quantum capacitance value without requiring complicated equipment has been experimentally investigated by modifying the Fermi level of large-area graphene using an external electric field. When the total capacitance change, caused by the gate bias in the passive graphene device, was increased to 60% compared to the initial state, a 6% shift in the resonant frequency could be achieved. While the signal characteristics of the graphene antenna are somewhat inferior compared to the conventional metal antenna, simplifying the device structure allowed reconfigurable characteristics to be implemented by using only the gate bias change.

Keywords: reconfigurable passive device; resonant frequency; graphene; capacitance change; frequency shift

1. Introduction

Graphene is a monolayer material but acts as an interface for bonding with other materials. Consequently, the primary area of study regarding the constant voltage condition has been the analysis of the resistance change of a device due to the interface charging phenomenon. For radio frequency (RF) applications, the inductance and capacitance of the device greatly affect the operating characteristics. Moreover, unlike metal, most signals applied from the outside would be transmitted, rather than absorbed or reflected, owing to graphene's thinness (theoretically ~0.3 nm) and possibly being due to the frequency dependence of skin depth. This suggests that the material with a thickness in nm is more reactive in the THz region [1–3].

Early studies regarding the RF characteristics of graphene devices primarily focused on applied research, using a high cutoff frequency and the frequency-doubling effect in field-effect transistor (FET)-type active devices [4–8]. Graphene FET (GFET)'s ground–signal–ground (GSG) pad structure, used for rapid signal propagation and amplification, was used for studying frequency ranges in the THz and optical domains. For these studies, exfoliated graphene or small-patterned chemical vapor deposition (CVD) graphene was used [4,9]. Passive RF devices have been investigated using graphene [10–23]. Most of the reports on passive devices using graphene, such as antennas and RF filters, analyzed them using theoretical approaches [20,21] because the size of graphene required to fabricate the passive devices was not available. For example, the size of an antenna resonating at 1 GHz requires a ~30 cm wire. For practical applications, GHz signals can be transmitted and received using shorter antennas by utilizing the multiple resonance phenomena and signal amplification techniques [11]. The drawback of this approach is the higher energy consumption required for signal processing. If a graphene antenna is too small compared to the wavelength, it cannot distinguish between ambient noise and signals. Thus, large-scale, high-quality graphene with an area of only a few millimeters is necessary for RF applications.

In this work, two dipole antennas integrated on the same plane were fabricated using a four-inch wafer-scale graphene sheet. When graphene was used as an antenna, a 2% level of resonance-frequency modulation was observed, according to the 10% change in capacitance. In addition, when the graphene was used as a ground plane, a 6% level of resonance-frequency modulation was successfully achieved due to the 60% capacitance change achieved via the quantum capacitance of the graphene sheet. For reconfigurable passive devices requiring resonant frequency control [24,25], a resonant frequency change rate corresponding to 2.5~9.6% of the capacitance change rate in the passive graphene device was possible.

2. Methodology: Structure of the Passive Graphene Device and Fabrication

To create the graphene-antenna device, a 50 nm oxide trench pattern was formed on a SiO_2 (300 nm)/Si substrate using a reactive ion-etching process. Subsequently, a 50 nm Pt electrode was deposited inside the trench by using an E-beam evaporation system to form a buried-gate electrode with significant advantages over the other structures in terms of the uniformity of the electric field applied to the channel region. To achieve planarization of the gate electrode, using chemical and mechanical polishing processes, 20 nm of Al_2O_3 was deposited using an atomic layer deposition (ALD) process and a 300 °C vacuum-annealing process for 1 h. Subsequently, the 4-inch-scale CVD graphene sheets (Alphagraphene Inc., Pohang, Republic of Korea) were transferred using a vacuum-transfer method to ensure a stable and consistent device operation, considering that this method can reduce impurities, such as oxygen and moisture, between the graphene and the substrate [26]. To prevent PR residues from appearing on the graphene channel, which may occur when patterning the graphene channel using the O_2 plasma, the graphene area was patterned via a photolithography process using a metal hard mask (Au). Finally, the top electrode (Au) was formed using an E-beam evaporation process and wet etching.

A passive device with a transmitter and receiver of two basic dipole antennas integrated on the same plane was fabricated to study the reconfigurable passive device using graphene. Device 1 consisted of 100 μm pitch ground–signal–ground (GSG) pads, as shown in Figure 1a, with a graphene antenna width of 50 μm and length of 450 μm. The thickness of the Al_2O_3 gate dielectric was 20 nm (dielectric constant = 6). The antenna itself was simply designed to operate at approximately 660 GHz, with resonance at 10–20 GHz. The ground plane was made of metal and platinum (Pt) to control the Fermi level of the graphene. The spacing between the transmitter and the receiver was alternated between 10 and 150 μm to observe the changes in signal-transmission characteristics. The Fermi level of the graphene could be controlled using the bias applied to the V_g contact shown in Figure 1a. Device 2, shown in Figure 1b, was a variation of Device 1 in which the ground plane was replaced with graphene. As the device capacitance was determined by the overlapping region of the metal and graphene, the total capacitance was expected to be similar to that of Device 1. As a reference, a similar device made of a metal (Au) film was fabricated, as shown in Figure 1c.

Its characteristics in the RF domain were analyzed using a network analyzer (Agilent 8510C) and a two-port probe system in the 1–40 GHz range. The S-parameters, reflections (S11, S22), and transmissions (S12, S21) were measured using a network analyzer. A constant voltage applied to the ground plane modulated the Fermi level of the graphene. Then, the S-parameters were measured as a function of the Fermi level of the graphene. First, we observed the appearance of signal reflections and transmissions when the spacing between the three antennas physically changed. The Fermi level of the graphene for the graphene antenna was assumed to be at the Dirac point. The observation of the Dirac point, based on the doping condition of graphene devices, can generally be obtained by measuring the I_d–V_g (drain current–gate voltage). In a passive device, such as Device 1, a DC current does not flow. Hence, the gate voltage value for accurate Dirac point setting was unknown during the RF measurements. Moreover, Dirac point analysis in graphene devices is possible using C–V measurements instead of I_d–V_g measurements. However, reconfiguring the four electrodes into one electrode for correct C–V measurements would be necessary, as two dipole antennas were patterned with graphene during the formation of Device 1. A capacitance measurement was carried out using the top electrode and buried electrode on the same substrate of the antenna device. Consequently, the Fermi level change in the graphene at the S-parameter measurement was set to a point where the Dirac point appeared at a gate voltage of 0 V. In this case, the estimated carrier density existing at the charge neutrality point was $3.1 \times 10^{11} cm^{-2}$. This value is very important because the quantum capacitance of graphene is generally more clearly observed when it has a lower carrier density than the classical capacitance of the device, and this is the ultimate reason for the tunability of the resonance frequency in these devices [27].

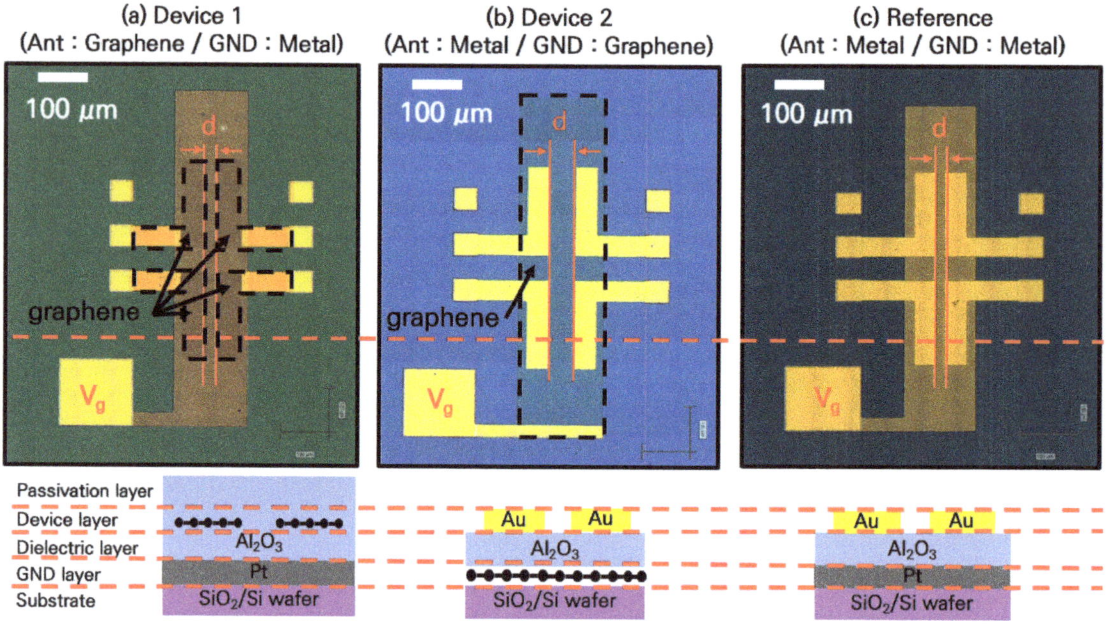

Figure 1. Optical images of the RF passive devices. Two dipole antennas are integrated on the same plane, and the ground plane is used as a gate to control the graphene Fermi level. (**a**) Device 1: has an antenna consisting of graphene patterns and ground metal. (**b**) Device 2: Graphene is used as the ground plane and the antenna is made of a metal film to increase the signal transmission rate of the device and induce a quantum capacitance change in graphene. (**c**) Reference device fabricated with Au patterns.

3. Results: RF Characteristics of the Reconfigurable Passive Graphene Device

Figure 2 shows the S-parameter values that were measured while alternating the antenna spacing between 10μm and 150μm. In the case of S11, the metal–metal structure showed a gradual decrease as the frequency increased. The graphene–metal structure showed a resonant peak at ~18 GHz. The inset of Figure 2a for the graphene–metal structure shows a gradual peak shift towards a higher frequency with wider gaps. On the other hand, the metal–graphene structure showed similar behavior as that of the metal–metal reference. As the distance between the antennas in Device 1 increased, the resonant frequency shifted towards a lower frequency and the overall transmittance decreased. The characteristics of the reference passive device were also measured under the same conditions. When the distance between the antennas was small, the resonance phenomenon was not clear, unlike with the graphene RF device. However, when the distance between the antennas was more than 100 μm, the resonant frequency became apparent. The reference device exhibited a higher transmittance value than Device 1 (by approximately 15 dB), but its reflectance value (S11) did not exhibit any resonance phenomena.

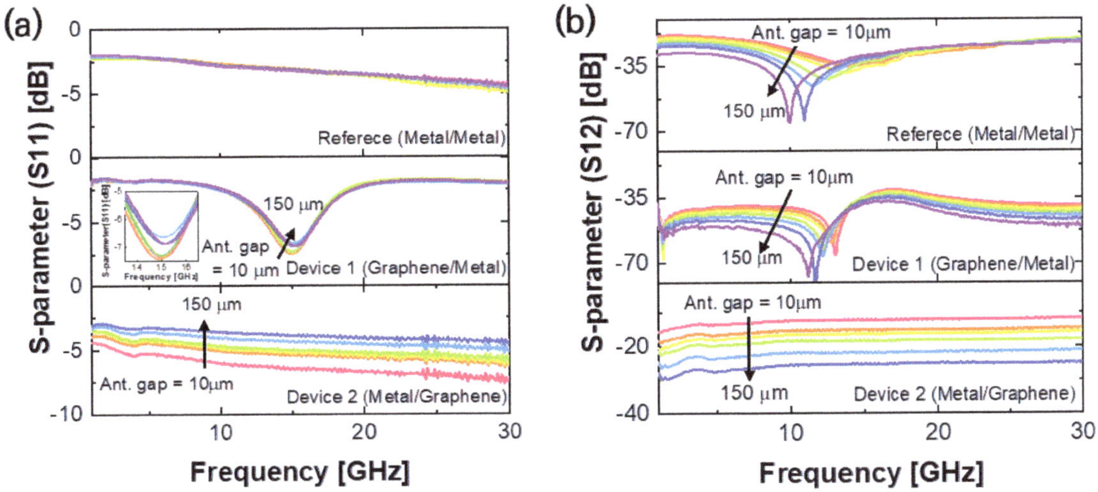

Figure 2. S-parameter measurement results based on the distance between the three antennas. (**a**) Reflectance (S11) and resonant frequency increase as the gap increases. (**b**) Transmittance (S12) and resonant frequency decrease as the gap increases.

In the case of S12, as shown in Figure 2b, as the antenna spacing increased, the resonant frequency shifted to lower frequencies and its value decreased, i.e., as the antenna spacing increased, signals of lower frequencies were transmitted better, while the signals of higher frequencies were mostly reflected at the input port. Consequently, as the distance between the antennas widened, the size of the ground plane increased, i.e., the capacitance of the entire device increased, and the high-frequency components were filtered.

The S-parameter measurements of the RF operating characteristics of Device 2 did not exhibit resonance at a specific frequency as they did in Device 1 or in the reference device. For Device 2, analysis was conducted based on the characteristic changes at 4.5 GHz. When the distance between the metal antennas was increased in Device 2, a change in the S-parameter value over the entire frequency range, rather than a resonant frequency shift, was observed. In particular, S12 or S21 showed signal transmission attenuations of more than 15 dB when the distance was increased. Simultaneously, in S11 or S22, the reflection amount increased as the distance between the antennas increased.

Next, the RF characteristics were investigated while applying the external bias to the graphene to modulate the Fermi level of the graphene. Using a gate electrode as a ground plane, measurements were made at −5, −3, −1, 0, 1, 3, and 5 V; different polarities were applied to observe the changes in the graphene carrier types in the graphene layer.

As resonance was most apparent when the antenna spacing was at 100 μm, the characteristics based on the gate bias change were observed in the device at that spacing. The results, based on changing the gate bias, did not differ much compared to those obtained when the antenna spacing was changed, as shown in Figure 3. As the value of the gate bias increased, the reflection decreased and the resonant frequency shifted to a lower frequency band. However, since there was no change to S11 over a frequency range of 15.5 GHz, it is difficult to appreciate that the resonant frequency shift is similar, i.e., the increase in the resonance bandwidth could be predicted. In S12, as shown in the Figure 3b inset, a parallel shift in the resonant frequency was observed, unlike in S11. As the V_g

increased, the resonant frequency shifted towards the lower frequencies. The frequency shift range was approximately 150 MHz. The change in characteristics due to the V_g was shown in that the degree of the shifting of the resonant frequency was proportional to the absolute value, regardless of the polarity of the V_g, i.e., when the Fermi level of graphene changed via the gate bias, the carrier type and carrier concentration changed based on this value and the polarity of the gate bias. However, as seen from the I_d–V_g measurements of the graphene device, the same conductance could be seen even if the absolute value of the gate bias was the same, regardless of the carrier type. Therefore, as shown in Figure 3, the change in resonant frequency in the passive graphene device was independent of the graphene carrier type but was related to the carrier concentration. As the voltage applied to regulate the graphene Fermi level was constant, the charge trap at the graphene interface in the device could be sensitive to certain carrier types and affect the device's performance.

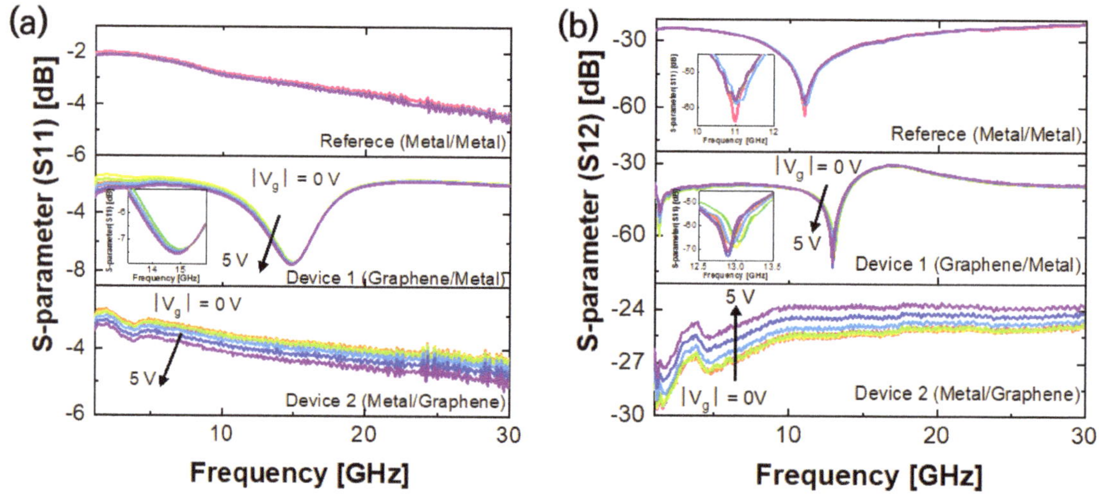

Figure 3. S-parameter measurement results with gate bias control. (**a**) Reflectance (S11) and (**b**) transmittance (S12).

Although Device 1 and the graphene RF device exhibited similar resonance characteristics to those of the reference passive device under its operating conditions, their transmittance values were low and their reflectance values were very high, possibly due to the significantly high contact resistance between the graphene and the graphene–metal contact. Therefore, improving the structure of the passive device would be necessary to secure its basic performance while facilitating the movement of resonant frequency by controlling the quantum capacitance of the graphene. The changes in the resonant frequency and measured S-parameter values due to changes in the Fermi levels of graphene, used as a ground plane by the gate bias, were more apparent than those in Device 1. As the V_g increased, the signal transmission improved by 3 dB and the reflectance reduced by 1 dB, as shown in Figure 3.

Under the same conditions, the measured values of Device 1, Device 2, and the reference device were compared to analyze the cause of different signal transmissions and reflection characteristics, as shown in Figure A1. For S11, the response of the reference device and Device 2 were similar when using metal antennas. For S12, the reference device and Device 1 exhibited similar responses when both used metal as the ground plane. Consequently, the reflection value of the device significantly influenced the impedance value of the antenna, and the transmission value significantly influenced the condition of the ground plane. The graphene antennas exhibited resonance at 15 GHz, but they have a very high reflection value in the remaining frequency range.

Conversely, in the case of signal transmission (S12, S21), the impedance value of the passive device itself and the space for transmitting the RF signal were important [25]. The RF signal was not transmitted through the surface of the metal, but rather through the space between the metal and ground plane such that a specific energy wavelength was emitted. Therefore, the state of the ground plane was an important aspect of the operating characteristics of the passive device. The problem arises from the graphene being used as an electrode, where the energy absorption rate of the electric field is low. However, graphene used as an electrode (ground) in actual passive devices has not been able to fulfill the role of a metal film. The electric field associated with graphene's Fermi level control is the same energy corresponding to this bias frequency. Therefore, examining how RF signals in the GHz range can be applied to graphene from a different perspective is necessary. Previous research has found that if graphene is placed in a floating state without any electrical contact, approximately 20% of the RF signal is absorbed and the remainder is reflected or transmitted [23]. Therefore, independent of the Fermi level control using a DC bias, 20% of the RF signal energy can be expected to be absorbed from the graphene ground plane, and the remaining part should be reflected, i.e., the RF signal will not be radiated properly between the antennas and the ground plane. In Device 1, this problem was eliminated as the RF signals were emitted from the graphene antennas, and the flow of the surface current was formed based on the graphene pattern.

Consequently, the degree of resonant frequency variation regarding the reflectance and transmittance due to the V_g was compared, as shown in Figure 4a,b. For the reference passive device, no resonant frequency change due to the V_g could be observed. However, a change in resonant frequency could be seen when the graphene was used. For S11, when Device 1 was exposed to a larger V_g value, an approximately 100 MHz shift towards the lower frequency was observed. However, for Device 2, the resonant frequency increased by approximately 200 MHz. The analysis results of Figure 4 show that the impedance of the antenna was affected more by a change in the S11 characteristics. An increase in the resonant frequency bandwidth indicated an improvement in signal transmission in the 15 GHz band. In Device 2, the resonant frequency shift and reflection decrease in the 4 GHz band could be attributed to an increase in carrier concentration caused by changes in Fermi levels. This resulted in improved signal transmission through the graphene ground plane. For S12, the shift width of the resonant frequency was more than that of S11, as shown in Figure 4. For both Device 1 and 2, the resonant frequency changed due to the V_g occurring approximately 1.5 times more, i.e., Device 1 and Device 2 exhibited a resonant frequency change of approximately 200 MHz and 300 MHz, respectively.

Increasing the carrier concentration of graphene in passive devices improved the RF signal transmission, and the quantum capacitance changes induced a change in the resonant frequency. The capacitance value of each passive device was measured to examine the correlation between the S-parameter characteristics and the capacitance changes in the passive device, as shown in Figure 4c. For Device 2 (Figure A2a), where graphene was used as the ground plane, a capacitance change of approximately 15 pF was measured due to the graphene quantum capacitance changes. Although the capacitance of the reference passive device (Figure A2b) was higher than that of Devices 1 and 2, the capacitance variation by the Vg was approximately 0.5 pF, which is approximately 30 times lower. Figure A2c compares the capacitance measurement results of Device 1, which used graphene as the antenna, and Device 2, which used graphene as the ground plane. The Dirac point not appearing near 0 V implies that the graphene was doped. As the Dirac point of Device 1 appears near −2.5 V, the graphene could be considered to be n-type. Since the Dirac point of Device 2 was measured by directly applying Vg to the graphene, the Fermi level shift was observed in the direction opposite to the graphene Fermi level shift in Device 1, i.e., as the Fermi level moved in the positive (p-type) voltage direction in Device 2, it mirrored the capacitance measurement value of Device 1. Figure 4c shows the rate of capacitance change. Owing to the changes in the quantum capacitance of the graphene, the capacitance value changed by approximately 120% as the Vg increased.

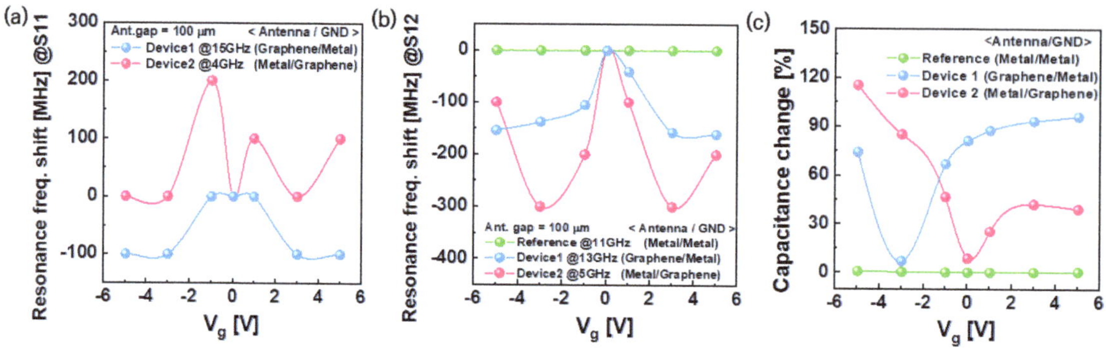

Figure 4. (**a**) For S11, the resonant frequency shift in reflectance is shown to be opposite when using graphene as an antenna and as a ground plane. As the graphene Fermi level increases in Device 1, the signal reflection in the low-frequency band increases. When the graphene Fermi level increases in Device 2, the signal reflection in the high-frequency band increases. Conversely, (**b**) for S12, all of the resonant frequency shifts were towards the lower frequency, exhibiting approximately twice the level of change in Device 1 compared to Device 2. (**c**) The capacitance variation according to the gate voltage change is plotted. When graphene is applied to the device, a capacitance change of approximately 120% is observed.

4. Discussion

In this experiment, we changed the resonant frequency in a passive device using graphene. The relationship between the rate of resonant frequency shift and the rate of capacitance change was investigated based on V_g = 0 V. As shown in Figure 5, 6% of the rate of transmittance resonant frequency shifted in correspondence with the rate of capacitance change that could be observed, regardless of the applied structure of graphene in the passive device. Additionally, the reflectance S-parameter value of the graphene RF devices changed minimally when compared with the reference device, and the rate of reflectance resonant frequency showed a 2% level shift, regardless of changes in the carrier concentration of graphene based on gate bias. Therefore, we can conclude that graphene RF devices can be reconfigured by gate bias without energy loss. When the total capacitance change of the passive device was 60% due to the graphene Fermi level change, a resonant frequency shift corresponding to 6% of the original resonant frequency value occurred.

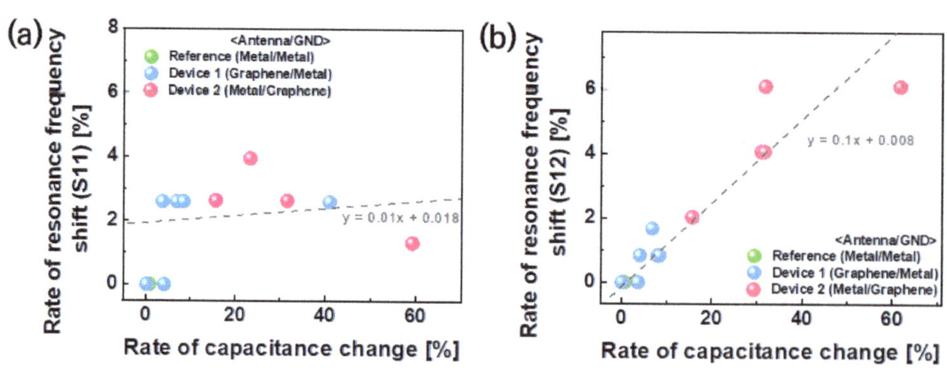

Figure 5. (**a**) Analysis of the relation between reflectance resonant frequency shift and graphene Fermi level change. (**b**) Analysis of the relation between transmittance resonant frequency shift and graphene Fermi level change.

5. Conclusions

To design a reconfigurable passive device using graphene, we fabricated integrated passive devices with two dipole antennas and analyzed their RF characteristics, aiming to analyze the use of graphene in passive devices and the amount of resonant frequency change possible based on the changes in V_g. When graphene was used as an antenna, characteristics similar to those of a passive device using a conventional metal film were observed, suggesting that graphene could be applied as a passive device based on the patterning. However, when a graphene ground plane was used, small-signal RF transmission was compromised and an accurate passive device operation could not be performed. Nonetheless, changes in quantum capacitance and carrier concentration due to graphene Fermi level changes affected the resonant frequency control and signal propagation. For reconfigurable passive devices requiring resonant frequency control, a resonant frequency change rate corresponding to 2.5~9.6% of the capacitance change rate in the passive graphene device was possible. Although these changes are insufficient for the practical application of graphene in reconfigurable passive devices, they indicate the potential to explore and positively contribute to their development.

Author Contributions: H.J.H.: investigation, conceptualization, methodology, validation, and writing—original draft; S.-Y.K. and S.K.L.: validation, resources, and data curation; B.H.L.: supervision and writing—review and editing. All authors have read and agreed to the published version of the manuscript.

Funding: This research received no external funding.

Data Availability Statement: Not applicable.

Acknowledgments: This research was supported by the FEOL Platform Development Project (Grant No. 2020M3F3A2A02082436), and by the Basic Science Research Program through the National Research Foundation of Korea (NRF) funded by the Ministry of Education (2020R1I1A1A01060352).

Conflicts of Interest: The authors declare no conflict of interest.

Appendix A

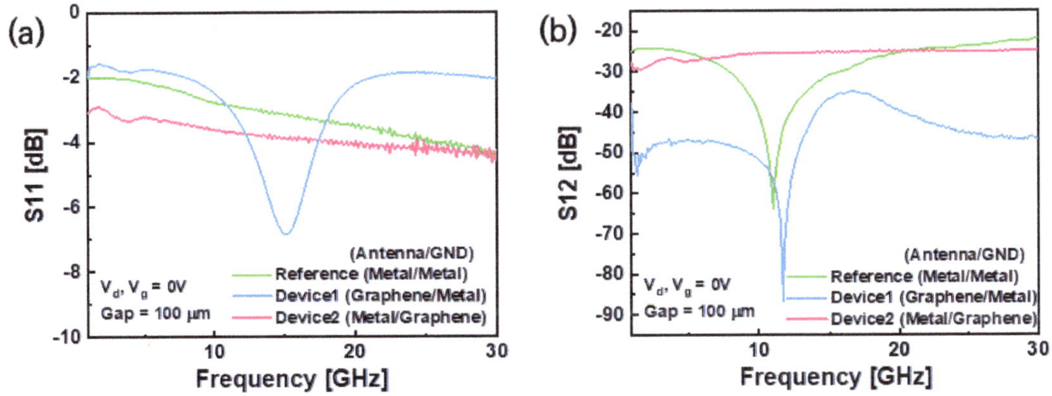

Figure A1. The measured values of Device 1, Device 2, and the reference device, under the same conditions, were compared to analyze the cause of different signal (**a**) reflection and (**b**) transmission characteristics.

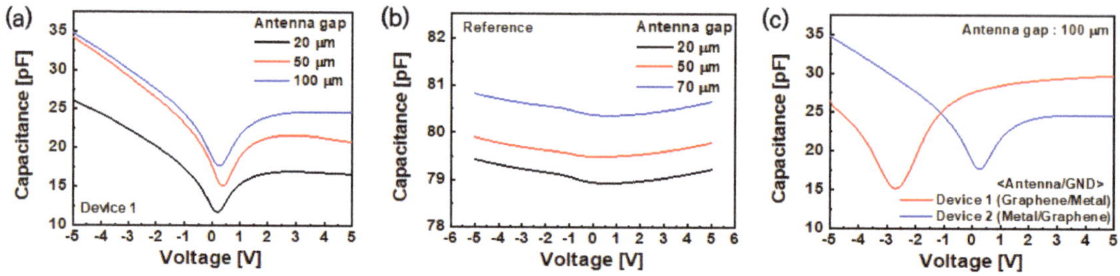

Figure A2. Comparison of capacitance variation of passive devices. (**a**) In Device 1, the capacitance value increases as the distance between the antennas increases. (**b**) For the reference device, there is no capacitance change due to quantum capacitance because graphene is not applied to the device. (**c**) Comparison of capacitance values of Device 1 and Device 2.

References

1. Schurr, J.; Ahlers, F.; Kibble, B.P. The ac quantum Hall resistance as an electrical impedance standard and its role in the SI. *Meas. Sci. Technol.* **2012**, *23*, 124009. [CrossRef]
2. Costa, K.Q.; Dmitriev, V.; Nascimento, C.M.; Silvano, G.L. Graphene Nanoantennas with Different Shapes. In Proceedings of the 2013 SBMO/IEEE MTT-S International Microwave & Optoelectronics Conference, Rio de Janeiro, Brazil, 4–7 August 2013. [CrossRef]
3. Dragoman, M.; Aldrigo, M. Graphene rectenna for efficient energy harvesting at terahertz frequencies. *Appl. Phys. Lett.* **2016**, *109*, 113105. [CrossRef]
4. Wang, H.; Nezich, D.; Kong, J.; Palacios, T. Graphene Frequency Multipliers. *IEEE Electron Device Lett.* **2009**, *30*, 547–549. [CrossRef]
5. Lin, Y.-M.; Dimitrakopoulos, C.; Jenkins, K.A.; Farmer, D.B.; Chiu, H.-Y.; Grill, A.; Avouris, P. 100-GHz Transistors from Wafer-Scale Epitaxial Graphene. *Science* **2010**, *327*, 662. [CrossRef]
6. Schwierz, F. Graphene transistors. *Nat. Nanotechnol.* **2010**, *5*, 487–496. [CrossRef] [PubMed]
7. Wang, H.; Hsu, A.; Wu, J.; Kong, J.; Palacios, T. Graphene-Based Ambipolar RF Mixers. *IEEE Electron Device Lett.* **2010**, *31*, 906–908. [CrossRef]
8. Wang, Z.; Zhang, Z.; Xu, H.; Ding, L.; Wang, S.; Peng, L.-M. A high performance top-gate graphene field-effect transistor based frequency doubler. *Appl. Phys. Lett.* **2010**, *96*, 173104. [CrossRef]
9. Jeon, D.-Y.; Lee, K.J.; Kim, M.; Kim, D.C.; Chung, H.-J.; Woo, Y.-S.; Seo, S. Radio-Frequency Electrical Characteristics of Single Layer Graphene. *Jpn. J. Appl. Phys.* **2009**, *48*, 091601. [CrossRef]
10. Deligeorgis, G.; Dragoman, M.; Neculoiu, D.; Konstantinidis, G.; Cismaru, A.; Plana, R. Microwave propagation in graphene *Appl. Phys. Lett.* **2009**, *95*, 073107. [CrossRef]
11. Novotny, L. Effective Wavelength Scaling for Optical Antennas. *Phys. Rev. Lett.* **2017**, *98*, 266802. [CrossRef]
12. Lin, Y.-M.; Jenkins, K.A.; Valdes-Garcia, A.; Small, J.P.; Farmer, D.B.; Avouris, P. Operation of Graphene Transistors at Gigahertz Frequencies. *Nano Lett.* **2009**, *9*, 422–426. [CrossRef] [PubMed]
13. Begliarbekov, M.; Strauf, C.P. Quantum inductance and high frequency oscillators in graphene nanoribbons. *Nanotechnology* **2010**, *22*, 165203. [CrossRef] [PubMed]
14. Dragoman, M.; Neculoiu, D.; Cismaru, A.; Muller, A.A.; Deligeorgis, G.; Konstantinidis, G.; Plana, R. Coplanar waveguide on graphene in the range 40 MHz–110 GHz. *Appl. Phys. Lett.* **2011**, *99*, 033112. [CrossRef]
15. Lee, H.-J.; Kim, E.; Jung, J. High Frequency Transmission Properties of Graphene Monolayer with Different Coplanar Waveguide Electrode Configurations. In Proceedings of the IEEE Nanotechnology Materials and Devices Conference, Jeju, Republic of Korea, 18–21 October 2011. [CrossRef]
16. Skulason, H.S.; Nguyen, H.V.; Guermoune, A.; Sridharan, V.; Siaj, M.; Caloz, C.; Szkopek, T. 110 GHz measurement of large-area graphene integrated in low-loss microwave structures. *Appl. Phys. Lett.* **2011**, *99*, 153504. [CrossRef]
17. Petrone, N.; Meric, I.; Chari, T.; Shepard, K.L.; Hone, J. Graphene Field-Effect Transistors for Radio-Frequency Flexible Electronics. *J. Electron Devices Soc.* **2015**, *3*, 44–48. [CrossRef]
18. Cheng, C.; Huang, B.; Liu, J.; Zhang, Z.; Mao, X.; Xue, P.; Chen, H. A Pure Frequency Tripler Based on CVD Graphene. *IEEE Electron Device Lett.* **2016**, *37*, 785. [CrossRef]
19. Kuzhir, P.P.; Paddubskaya, A.G.; Volynets, N.I.; Batrakov, K.; Kaplas, T.; Lamberti, P.; Kotsilkova, R.; Lambin, P. Main principles of passive devices based on graphene and carbon films in microwave—THz frequency range. *J. Nanophotonics* **2017**, *11*, 032504. [CrossRef]
20. Krupka, J.; Strupinski, W.; Kwietniewski, N. Microwave Conductivity of Very Thin Graphene and Metal Films. *J. Nanosci. Nanotechnol.* **2011**, *11*, 3358–3362. [CrossRef]

21. Aldrigo, M.; Dragoman, M.; Costanzo, A.; Dragoman, D. Graphene as a high impedance surface for ultra-wideband electromagnetic waves. *J. Appl. Phys.* **2013**, *114*, 184308. [CrossRef]
22. Shangguan, Q.; Zhao, Y.; Song, Z.; Wang, J.; Yang, H.; Chen, J.; Liu, C.; Cheng, S.; Yang, W.; Yi, Z. High sensitivity active adjustable graphene absorber for refractive index sensing applications. *Diam. Relat. Mater.* **2022**, *128*, 109273. [CrossRef]
23. Shangguan, Q.; Chen, Z.; Yang, H.; Cheng, S.; Yang, W.; Yi, Z.; Wu, X.; Wang, S.; Yi, Y.; Wu, P. Design of ultra-narrow band graphene refractive index sensor. *Sensors* **2022**, *22*, 6483. [CrossRef] [PubMed]
24. Yang, C.-L.; Li, C.-S. Reconfigurable Antennas of Wide Tuning Ranges and Controllable Selectivity Using Matching Networks (Chapter 17). In *Ultra Wideband Communications: Novel Trends—Antennas and Propagation*; INTECH Europe: Rijeka, Croatia, 2011. [CrossRef]
25. Rodrigo, D.; Cetiner, B.A.; Jofre, L. Frequency, Radiation Pattern and Polarization Reconfigurable Antenna using a Parasitic Pixel Layer. *IEEE Trans. Antennas Propag.* **2014**, *62*, 3422–3427. [CrossRef]
26. Lee, S.; Lee, S.K.; Kang, C.G.; Cho, C.; Lee, Y.G.; Jung, U.; Lee, B.H. Graphene transfer in vacuum yielding high quality interface. *Carbon* **2015**, *93*, 286–294. [CrossRef]
27. Xia, J.; Chen, F.; Li, J.; Tao, N. Measurement of the quantum capacitance of graphene. *Nat. Nanotechnol.* **2009**, *4*, 505–509. [CrossRef] [PubMed]

nanomaterials

MDPI

Article

Enlarged Interlayer Spacing of Marigold-Shaped 1T-MoS_2 with Sulfur Vacancies via Oxygen-Assisted Phosphorus Embedding for Rechargeable Zinc-Ion Batteries

Qinhu Xu, Xinyu Li *, Luchen Wu, Zhen Zhang, Yong Chen, Ling Liu and Yong Cheng *

College of Science & Ministry-Province Jointly-Constructed Cultivation Base for State Key Laboratory of Processing for Mom-Ferrous Metal and Featured Materials & Key Lab. of Nonferrous Materials and New Processing Technology & Network and Information Center, Guilin University of Technology, Guilin 541004, China

* Correspondence: lixinyu5260@163.com (X.L.); hb_cy@163.com (Y.C.)

Abstract: Structural unsteadiness and sluggish diffusion of divalent zinc cations in cathodes during cycling severely limit further applications of MoS_2 for rechargeable aqueous zinc-ion batteries (ZIBs). To circumvent these hurdles, herein, phosphorus (P) atom embedded three-dimensional marigold-shaped 1T MoS_2 structures combined with the design of S vacancies (Sv) are synthesized via the oxygen-assisted solvent heat method. The oxygen-assisted method is utilized to aid the P-embedding into the MoS_2 crystal, which can expand the interlayer spacing of P-MoS_2 and strengthen Zn^{2+} intercalation/deintercalation. Meanwhile, the three-dimensional marigold-shaped structure with 1T phase retains the internal free space, can adapt to the volume change during charge and discharge, and improve the overall conductivity. Moreover, Sv is not only conducive to the formation of rich active sites to diffuse electrons and Zn^{2+} but also improves the storage capacity of Zn^{2+}. The electrochemical results show that P-MoS_2 can reach a high specific capacity of 249 mAh g^{-1} at 0.1 A g^{-1}. The capacity remains at 102 mAh g^{-1} after 3260 cycles at a current of 0.5 A g^{-1}, showing excellent electrochemical performance for Zn^{2+} ion storage. This research provides a more efficient method of P atom embedded MoS_2-based electrodes and will heighten our comprehension of developing cathodes for the ZIBs.

Keywords: interlayer-expanded MoS_2; sulfur vacancies; phosphorus embedding; aqueous Zn-ion batteries

1. Introduction

Due to the distinct characteristics of zinc metal, such as high theoretical capacity (820 mAh g^{-1}), low redox potential (-0.76 V concerning standard hydrogen electrodes), non-toxicity, low cost, and intrinsic safety, as well as high reversibility in aqueous electrolytes, ZIBs hold promising potential for large-scale energy storage [1–4]. However, Zn^{2+} in aqueous solution is hard to intercalate between layers because it exists as hydrated zinc ions ($[Zn(H_2O)_6]^{2+}$) with a size of 0.404 nm~0.43 nm, making ($[Zn(H_2O)_6]^{2+}$) intercalation and deintercalation put forward higher requirements on the interlayer spacing and other properties of the cathode material [5–7]. In addition, divalent Zn^{2+} usually exhibits strong electrostatic interactions with the host lattice, inhibiting the diffusion process of Zn^{2+}, resulting in few cathode materials available for ZIBs, which are mainly confined to manganese-based materials, vanadium-based materials, and Prussian blue [8–10]. Therefore, the sluggish intercalation dynamics of divalent Zn^{2+} enable the search for appropriate cathode materials to be very important for the development and application of ZIBs.

Molybdenum disulfide (MoS_2) is a typical layered transition metal sulfide, and the covalently bonded S-Mo-S layers are stacked together by weak van der Waals forces, which is conducive to the intercalation of foreign guests [11–14]. Unfortunately, due to the strong electrostatic interaction between inserted Zn^{2+} with a large hydrated radius (4.3 Å) and host structures, Zn^{2+} is challenging to intercalate/deintercalate from the host frame, greatly affecting the reversible capability and rate characteristic of MoS_2 as cathodes arousing from its smaller layer spacing and the large inert base with fewer active sites [15–18]. Moreover, MoS_2 also has problems such as poor conductivity, poor hydrophilicity, agglomeration, and volume expansion, which limits the application of MoS_2 in ZIBs. Therefore, some pioneering efforts, including introducing phase transition engineering and tailoring nanostructures, have been made to improve the Zn^{2+} storage capability of MoS_2 cathodes [19–22]. One available strategy to resolve such challenges is to intercalate foreign elements into MoS_2 to enlarge layer spacing [15,23,24]. Experimental and calculation results indicate that P dopants could not only modify the surface electronic state of MoS_2 and increase its inherent conductivity but also lead to MoS_2 expansion and induce a partial phase transition of MoS_2 from hydrophobic (2H phase) to hydrophilic (1T phase); the 1T-MoS_2 has lower Zn diffusion energy barriers [25]. However, due to the spontaneous formation of MoP, resulting in P atom doping is inherently difficult and has lower P content [26,27]. Despite substantial achievements having been made, P-embedding 1T-MoS_2 to achieve extended layer spacing for enhancing the ability of Zn^{2+} intercalates/deintercalates is very necessary, thus requiring a facile method.

Additionally, vacancy engineering has been in the spotlight as a feasible strategy to enhance the zinc ions storage capability considerably. Especially based on previous reports [28], sulfur vacancies can provide abundant active sites as additional ion storage sites for Zn^{2+} as well as offer a fast electrochemical response. For example, Hu et al. [29] developed a few-layered MoS_2 anchored on an N-doped carbon flower with Sv as anode material for sodium-ion batteries, exhibiting excellent performance. Wang et al. [30] improved the performance of hydrogen evolution reaction (HER) by regulating the S-vacancy distribution and concentration in MoS_2. Xu et al. reported that the preparation of defect-rich MoS_2 accelerated the diffusion kinetics of Zn^{2+} to the active center [31]. Therefore, the rational S vacancy modulation strategy is essential to improve the reaction kinetics of Zn^{2+} for realizing its application in ZIBs.

Herein, an oxygen-assisted(O) strategy is used to aid the P-embedding into the MoS_2 crystal; the MoS_2 layer spacing is expanded, and the marigold-shaped 1T-MoS_2 nanosheets with rich S vacancies are prepared. In this process, oxygen played a key role in aiding the successful embedding of P into the lattice of 1T-MoS_2. The intercalation of P in MoS_2 achieves the desired effect, significantly enlarging MoS_2 interlayers (from 0.62 to 0.98 nm) and enhancing hydrophilicity. In addition, these S vacancies defects as active sites can make it easier for Zn^{2+} adsorption and desorption. The synergistic effect of P-embedding induced and extended the marigold-shaped 1T-MoS_2 layer spacing and caused the Sv to form an active center in the basal plane of MoS_2, effectively reducing the Zn^{2+} diffusion resistance. It provides an easier channel for the insertion of $[Zn(H_2O)_6]^{2+}$, resulting in rapid reaction kinetics. As expected, the P-MoS_2 electrode achieves a remarkably high capability of 249 mAh g^{-1} at 0.1 A g^{-1}, which is five times higher than the specific capacity of pristine MoS_2. In comparison to the pristine P-MoS_2 counterpart with minimal capacity delivery, P-MoS_2 can achieve a high reversible capacity of 105 mAh g^{-1} at 0.5 A g^{-1} with 3260 cycles and excellent capacity retention of 70 mAh g^{-1} at 1 A g^{-1}. This study will provide more efficient avenues for investigating more electrode materials with poor intercalation kinetics in ZIBs.

2. Methods

2.1. Experimental Section

Synthesis of P-MoS_2: In a representative procedure, 1 mmol (1236 mg) ammonium molybdate tetrahydrate ($(NH_4)_6Mo_7O_{24}\cdot4H_2O$, AR) and 30 mmol (2284 mg) thiourea (CH_4N_2S, AR) were added into 35 mL deionized water on the basis of a beforehand report with slight adjustment [32]. After being stirred for 30 min, an amount of sodium hypophosphite monohydrate ($NaH_2PO_2\cdot4H_2O$, AR, 400 mg) was dissolved into a mixed solution with magnetic stirring for 2 h. Then, the precursor solution was heated to 180 °C for 24 h in a 50 mL Teflon-lined stainless steel autoclave. After cooling the reaction system to air temperature, which was collected by centrifugation, cleaned multiple times with water, and then dried at 80 °C. As a contrast, pristine MoS_2 was also prepared without $NaH_2PO_2{\cdot}4H_2O$ following a similar synthetic route.

2.2. Material Characterizations

The morphology of materials was evaluated with a scanning electron microscope (SEM, Hitachi S-4800, Tokyo, Japan) and a transmission electron microscope (TEM, Tecnai G2 20 TWIN, FEI, Hillsboro, OR, USA). An energy dispersive spectrometer (EDS S-00123, USA) connected to the SEM was used to study the elemental composition and elemental analyses of the composites. At −196 °C, N_2 adsorption/desorption isotherms were measured with an automated SSA and porosity analyzer (asap2460). Brunauer–Emmett–Teller (BET) and Barrett–Joyner–Halenda (BJH) adsorption techniques were used to determine the SSA and mesopore size distribution of each sample. All samples' crystal phases and compositions were determined using X-ray diffraction (XRD, MiniFlex-600, Rigaku, Tokyo, Japan) and X-ray photoelectron spectroscopy (XPS, Thermo Scientific K-Alpha, Waltham, MA, USA). Raman spectra at 532 nm were measured with a Horiba Scientific LabRAM HR Evolution Raman spectrometer. The TGA/DTA was tested by thermogravimetric analysis equipment (SDT Q600, USA) from room temperature to 750 °C at a heating rate of 10 °C/min. A hydrophilicity test was measured using a contact angle test device (Dataphysics OCA20, Filderstadt, Germany).

2.3. Electrochemical Characterization

Zn served as the anode, glass fiber membrane served as diaphragms, and 3 M $Zn(CF_3SO_3)_2$ served as the electrolyte in CR2016-type coin cells used to investigate the Zn^{2+} storage properties of P-MoS_2. To manufacture the working electrode, P-MoS_2 (70 weight percent), super p carbon (20 weight percent), and polyvinylidene fluoride (10 weight percent) were thoroughly blended in *N*-methyl-2-pyrrolidone for 15 min. Finally, the aforementioned slurry was distributed over a clean stainless steel mesh and dried for 24 h at 60 °C. On a Neware battery tester (CT4008), galvanostatic charge/discharge measurements and galvanostatic intermittent titration technique (GITT) experiments were carried out continuously between 0.25 and 1.25 V (vs. Zn/Zn^{2+}). An electrochemical workstation was used for the cyclic voltammetry (CV) and electrochemical impedance spectroscopy (EIS) experiments (CHI-660D). When the batteries were completely charged, impedance measurements were taken (zinc extraction). At room temperature, all tests were conducteddw.

3. Results and Discussion

3.1. Composition and Structure

Figure 1 depicts the synthesis process of P-MoS_2 and pristine MoS_2. The phosphorus embedded three-dimensional marigold 1T MoS_2 was obtained by controlling the crystallization process. In a nutshell, an oxygen-rich atmosphere plays a vital role in decreasing the formation energy of P-embedding in MoS_2 [26]. The reaction process becomes more inadequate as the synthesis temperature decreases, resulting in the leftover oxygen inherited from the molybdate precursor, thus realizing the oxygen-rich atmosphere [32]. Additionally, most of the Mo^{4+} ions form ionic bonds with S^{2-}, which are self-assembled in the form of nanocrystals connected into three-dimensional marigold-shaped structures. Excess thiourea could be adsorbed on the surfaces of initial nanocrystallites and impede the formation of orientated crystals, which results in abundant Sv being produced amid the flat structure [33]. The SEM and TEM were utilized to explore the morphology evolution of as-prepared P-MoS_2 further. As shown in Figure 2a,b, the SEM of P-MoS_2 displays the magnificent marigold structure with an internal diameter of almost 1 μm, which was self-assembled from curved thickness nanosheets with transverse dimensions of 400–500 nm. In comparison, pristine MoS_2 has a rather formidable nanosheet microstructure, which is much larger in transverse size (diameter almost 4–5 μm) and displays severe aggregation in the absence of a well-tuned morphology (Figure S1). The relatively smaller size of the beautiful marigold structure should help to shorten the Zn^{2+} diffusion duration. The uneven nanostructure of P-MoS_2 with evident ripples and bending wrinkles is depicted in TEM images (Figure 2c,d), in which nanosheets are haphazardly joined and formed into three-dimensional structures. The TEM images in Figure 1e,f reveal a distinct lattice structure with some disorder and abundance of Sv (Figure S2), resulting in an incomplete lattice that may accommodate a host of unsaturated S atoms as active sites. Furthermore, the TEM images reveal that dozens of pristine MoS_2 layers are heavily stacked together with an interlayer spacing of 0.62 nm (Figure S3), whereas the stacking of the P-MoS_2 layers is significantly relieved with the interlayer distance strikingly expanding to 0.98 (Figure S4) and 0.86 nm. The results show that P-embedding can enlarge the layer spacing of P-MoS_2, which is advantageous to strengthen Zn^{2+} intercalates/deintercalates. Figure 2h–k displays the mapping images of P-MoS_2, which expressly demonstrate that Mo, O, S, and P elements are distributed uniformly. Meanwhile, Figure S5 shows the elemental content of P-MoS_2.

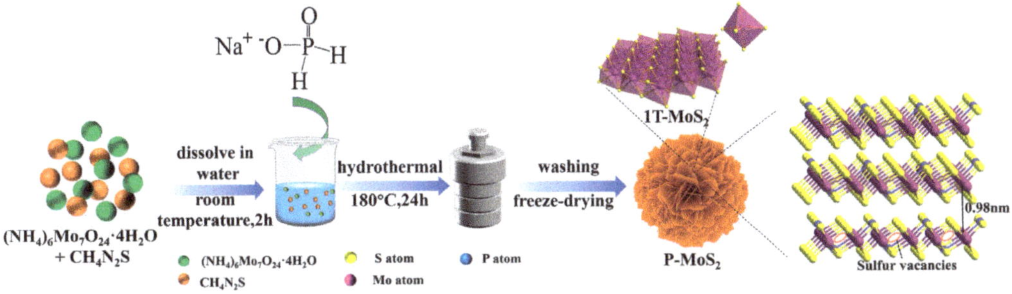

Figure 1. Schematic illustration of synthesis P-MoS_2.

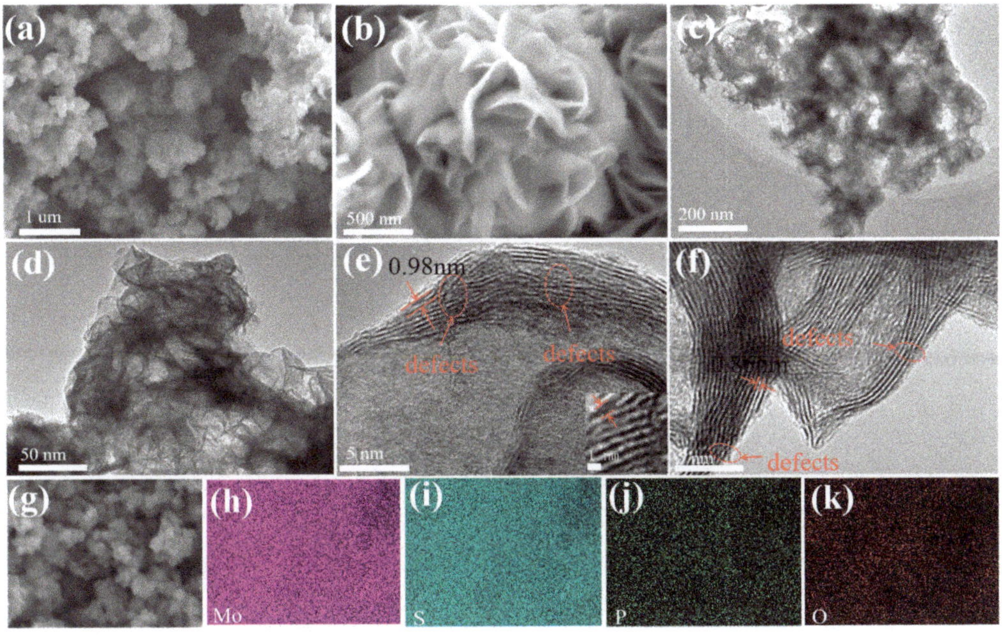

Figure 2. (**a**,**b**) SEM; (**c**,**d**) TEM; (**e**,**f**) HRTEM; (**g**–**k**) EDS of P-MoS_2.

The structural information of the two obtained samples was studied by X-ray diffraction (XRD) analysis (Figure 3a). Compared with the highly crystalline pristine MoS_2 in the 2H-MoS_2 phase (JCPDS card number 37–1492), all peaks of the P-MoS_2 are broadened due to nanoscale effects and crystal structural defects. The weakening of the (002) diffraction peak of P-MoS_2 indicates a low stacking height along this direction [34,35]. Meanwhile, the (002) peak of P-MoS_2 slightly shifts to a lower angle (14.2° → 13.96°), indicating the formation of sulfur vacancies in MoS_2 [36,37]. In addition, a new peak appeared at 9.8°, which is related to the presence of P, leading to the formation of a stacking layer, thus extending the c-axis of MoS_2. According to Bragg's law, the lattice spacing of the stacking layer at 9.8° is consistent with the spacing in TEM. Due to P's larger atomic radius than S, P atoms were introduced into the MoS_2 matrix, causing the lattice to expand and contributing to the widening of interlayer spacing [25,38,39]. Figure 3b exhibits the Raman spectra of P-MoS_2. Three typical Raman scattering peaks at 280 cm^{-1} ($E^1{}_g$), 234 (J_2) cm^{-1}, and 334 cm^{-1} (J_3) are attributed to the octahedral coordination of metal 1T-MoS_2 [40–42]. P-MoS_2 exhibits two typical characteristic peaks at 376 cm^{-1} and 394 cm^{-1}, corresponding to the in-plane ($E^1{}_{2g}$) and out-of-plane ($A^1{}_g$) modes of 2H MoS_2, which are clearly distinct from pristine MoS_2. The distance between the $E^1{}_{2g}$ and $A^1{}_g$ peaks is 18 cm^{-1} for P-MoS_2 and 23 cm^{-1} for bulk MoS_2. Thus, the moving peak of $A^1{}_g$ might represent the substantial out-of-plane vibration, indicating that the decreased van der Waals force, together with the enlarged interlayer spacing of MoS_2 caused by P-embedding, is favourable for Zn^{2+} movement and storage [5,43]. The weaker intensity of the $E^1{}_{2g}$ peak in P-MoS_2 is weaker compared to other samples, thereby proving the presence of Sv in the crystal structure [44].

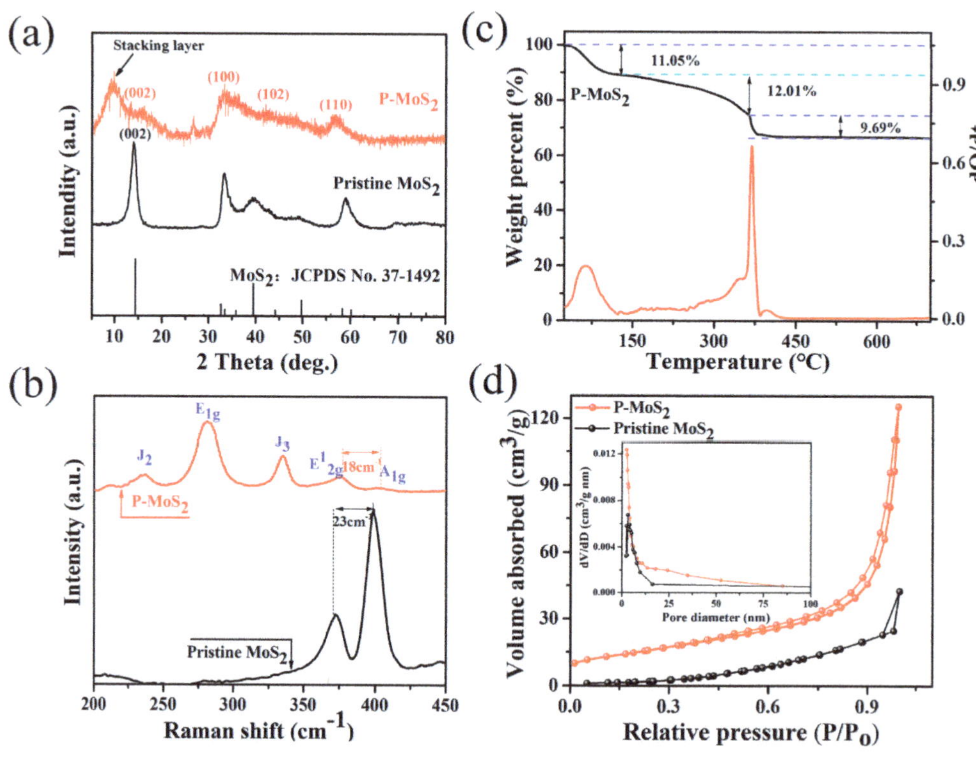

Figure 3. (**a**) XRD patterns of P-MoS_2 and pristine P-MoS_2. (**b**) Raman spectra of P-MoS_2 and pristine MoS_2. (**c**) TGA and DTA curves of P-MoS_2. (**d**) N_2 adsorption/desorption isotherm and corresponding pore size distribution of P-MoS_2 and pristine MoS_2.

The TGA-DTA results for the P-MoS_2 shown in Figure 3c show the initial mass loss of 11.05 wt% at 135 °C due to surface-adsorbed water evaporation. The mass loss of 12.01 wt% between 135 and 350 °C corresponds to the water loss integrated into the crystal structure, implying that the enlarged layer spacing of P-MoS_2 is probably caused by water insertion [45]. The additional weight loss of 9.69% over 350 °C may be caused by O atoms, which are utilized to aid P atoms embedded in the MoS_2 structure, with the unsaturated S atoms resulting In the reaction $2MoS_2 + 7O_2 \rightarrow 2MoO_3 + 4SO_2$ [46,47]. The first derivative curve of the related DTA curve shows a visible endothermic peak at 350 °C. This peak is caused by the transformation of the crystal structure of MoS_2, which is confirmed by XRD analysis. Moreover, the cumulative pore volume and BET surface area of P-MoS_2 and pristine MoS_2 were also measured. As shown in Figure 3d, P-MoS_2 and pristine MoS_2 exhibited typical type IV isotherms with hysteresis loops, indicating mesopores between MoS_2 nanosheets. Nevertheless, the hysteresis of P-MoS_2 starts at a lower pressure region ($P/P_0 \approx 0.5$) than that of pristine MoS_2 ($P/P_0 \approx 1$), showing their different porous nature. Furthermore, the wide hysteresis of P-MoS_2 indicates its increase in porosity. According to the pore volume–pore size distribution curve, P-MoS_2 has more abundant mesoporous structures, which is conducive to electron transport. The increased specific surface area of P-MoS_2 is 51.572 $m^2\ g^{-1}$, which is significantly more than pristine MoS_2 (11 $m^2\ g^{-1}$). The increased specific surface area improves the interaction between the material's exposed active sites and the electrolyte, which favors increasing the number of active sites for Zn^{2+} storage. These results clearly show that the P-embedding strategy can greatly increase the surface area and pore volume of MoS_2. To evaluate the hydrophilicity of P-MoS_2, we performed a contact angle test. Water contact angles of P-MoS_2 drop from 54.46° to 38.8° (Figure S6), suggesting improved hydrophilicity and beneficial to the diffusion of Zn^{2+} [6].

X-ray photoelectron spectroscopy (XPS) reveals detailed valence states and chemical contents of manufactured materials. The survey XPS spectra for P-MoS_2 and is displayed in Figure S7. As shown in Figure 4a, the XPS spectra of the Mo 3d scan contain two sets of doublet peaks (228.55 and 231.76 eV; 229.5 and 232.75 eV), which belong to Mo $3d_{5/2}$ and Mo $3d_{3/2}$ of Mo^{4+} for 1T-MoS_2 and 2H-MoS_2 [19,48,49]. This suggests the coexistence of 2H and 1T in P-MoS_2, which is compatible with Raman data. Similarly, differences can be observed in the S 2p spectra in Figure 4b, where peaks at 161.4 (S $3d_{3/2}$ of S^{2-}) and 162.7 eV (S $3d_{1/2}$ of S^{2-}) attributed to 1T MoS_2, while peaks at 162.0 and 163.1 eV ascribed to 2H-MoS_2 [49]. After fitting S 2p spectrum, an extra peak can be seen at 164.19 eV, according to the relevant literature, ascribed to the edge S [50]. Furthermore, using deconvolutions of Mo 3d, the proportion of 1T is estimated at 53% (Table S1), which is equivalent to the product obtained from chemical exfoliation [21]. Meanwhile, the existence of the O 1s signal offers conclusive proof of oxygen-assisted P-embedding in P-MoS_2 (Figure 4c). It can be deconvoluted into three peaks at 530.42, 531.94, and 533.44 eV, which is ascribed to the P-O bond, Mo-O bond, and adsorbed water, respectively. The analysis shows that oxygen plays an auxiliary role in P-embedding and can also make P-MoS_2 form a three-dimensional beautiful marigold structure in a bonding manner. P 2p signals were detected at 134.02, 130.99, and 130.06 eV (Figure 4d), confirming the existence of P. The dominant P signal at 134.6 eV can be allocated to the ${PO_4}^{3-}$ species, whereas the remaining two should be attributed to Mo-P, indicating that P atoms are embedded in the MoS_2 lattice [44,51]. Furthermore, elemental composition analysis reveals that the Mo/S ratio is around 1:1.73 (Table S2), which is much lower than the 1:1.95 for MoS_2 by XPS. This finding implies that P-MoS_2 has a substantial number of S vacancies [5,20,45].

3.2. Electrochemical Performance of Aqueous Zn-Ion Batteries and Kinetic

To investigate the Zn^{2+} storage capacity of the prepared samples as stand-alone cathodes for AZIBs, CR2016 coin cells were manufactured in the air environment (see the experimental section for details). Cyclic voltammetry (CV) curves in a voltage window of 0.25 to 1.25 V of 0.1 mV s^{-1} are shown in Figure 5a,b. In contrast to pristine MoS_2, P-MoS_2 has two redox peaks: the cathodic peak at 0.65 V linked to Zn^{2+} insertion (possibly overlapping or merging with the reduction in Mo^{6+}/Mo^{4+}) and an anodic peak at 0.98 V is associated to Zn^{2+} inlay removal [23]. Furthermore, the second and third CV cycles of the two electrodes practically coincide, showing high reversibility. Surprisingly, the peak area of P-MoS_2 is considerably larger than that of pristine MoS_2, showing that the enlarged layer spacing and Sv can significantly boost volume capacity. The P-MoS_2 charge/discharge (CD) curves are comparable with the CV data (Figure 5c), which shows an intercalation plateau of P-MoS_2 approximately 0.6 V. In addition, the specific capacity of P-MoS_2 is 249 mA h g^{-1}, which is more than five times that of pristine MoS_2 (50 mA h g^{-1}), significantly better than the well-known Zn^{2+} intercalation host, i.e., the Chevrel Phase Mo_6S_8 (60 mA h g^{-1} at 0.06 A g^{-1}) [52,53]. Meanwhile, the capacity of several modified MoS_2 materials is compared at 0.1 A g^{-1} in Table S3. Figure 5d shows display ratio capability with the volumes of P-MoS_2, which are 249, 158, 125, 97, and 75 mAh g^{-1} at 0.1, 0.2, 0.5, 1, and 2 A g^{-1}, respectively, and Figure 5e shows the charge/discharge curves of the first cycle capacity. P-MoS_2 capacitance achieves 143 mAh g^{-1} (91% capacity retention) when it reaches 0.2 A g^{-1}. These hint at the excellent rate characteristic and rapid dynamics of P-MoS_2. In contrast, pristine MoS_2 offers rather low capacities (0–45 mAh g^{-1}) at a variety of current densities. P-MoS_2 demonstrates a consistent capacity of 70 mAh g^{-1} after 200 cycles at 1 A g^{-1} to further illustrate its outstanding working life (Figure 5f). The stability was further investigated at 0.5 A g^{-1} in Figure 5g.

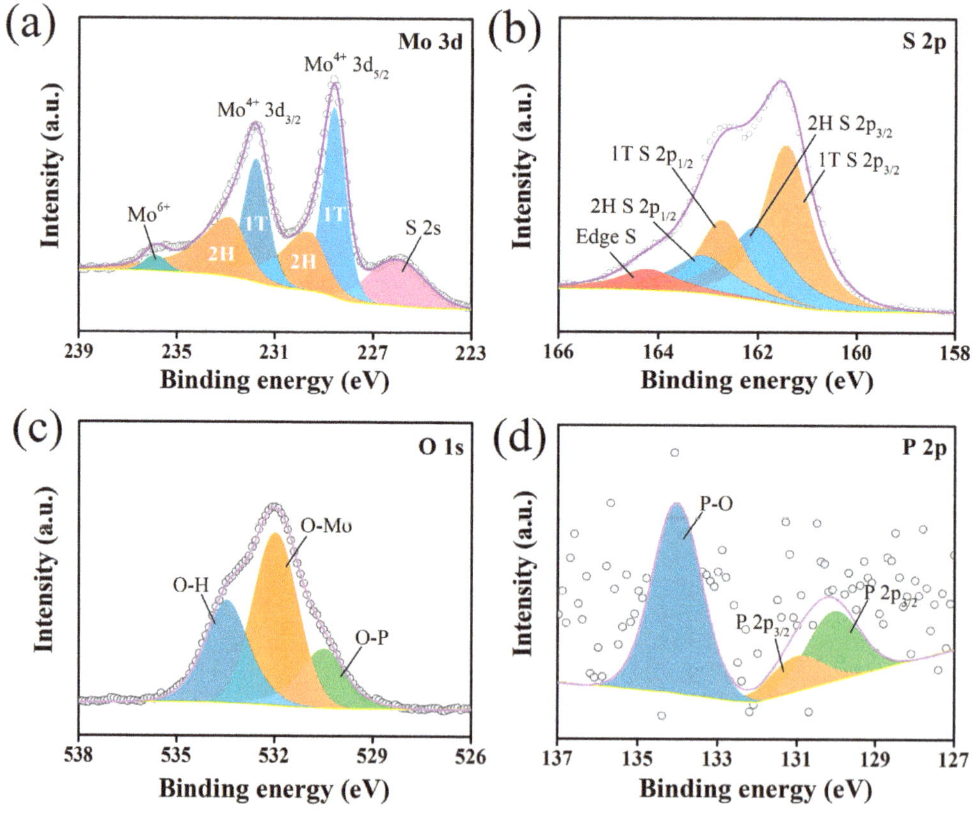

Figure 4. XPS spectra of P-MoS_2: (**a**) Mo 3d, (**b**) S 2p, (**c**) O 1s, and (**d**) P 2p.

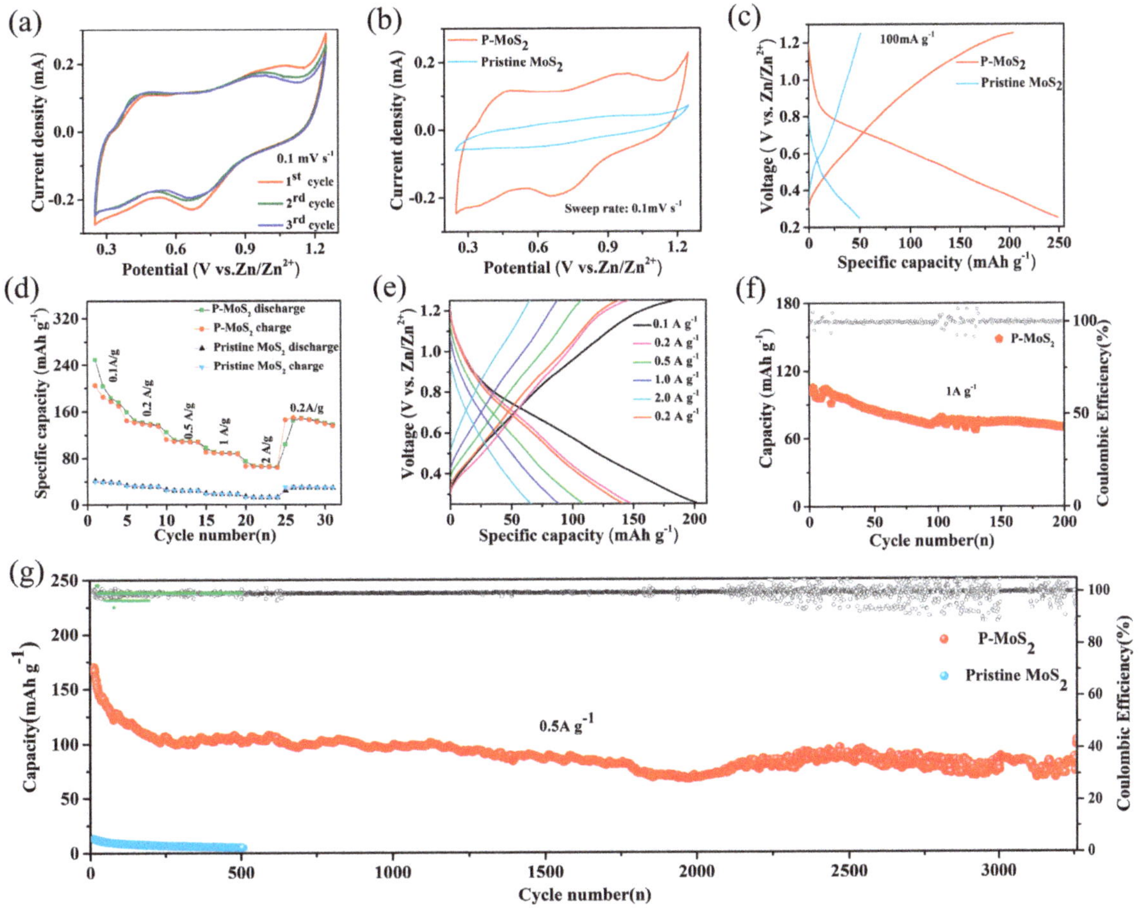

Figure 5. Electrochemical performance of P-MoS_2 and pristine MoS_2 (**a**) CV curves at 0.1 mV s^{-1}. (**b**) CV at 0.1 mV s^{-1} rate of P-MoS_2 and Pristine MoS_2. (**c**) Discharge/charge curves at 0.1 A g^{-1}. (**d**) Rate performance. (**e**) First cycle capacity at different cycles. (**f**) Two hundred cycling stability at 1 A g^{-1}. (**g**) Long-term cycling stability at 0.5 A g^{-1}.

P-MoS_2 maintains a capacity of 102 mAh g^{-1} after 3260 cycles, achieving capacity retention of 70%, with the efficiency of the electrode being close to 100%. However, the cycling stability of the P-MoS_2 electrode seems rather poor after 2000 cycles, which is attributed to the irreversible structural damage, volume changes, and unstable 1T phase during the continuous charging and discharging process [54,55]. In comparison, pristine MoS_2 has almost zero capacity, which is related to its poor conductivity and unstable structure [16,23].

The results show that P-MoS_2 outperforms pristine MoS_2 in terms of reversible capacity, cycle stability, and rate capacity. To better understand the electrochemical reaction behavior of the P-MoS_2 electrode, CV curves at varied scan speeds (0.1–1.0 mV s^{-1}) are tested (Figures 6a and S8). The oxidation peak gradually shifted toward high potential with an increasing scan rate, and the reduction peak shifted toward low potential because the electrode polarization broadened at a high scan rate. Whether pseudocapacitive behavior is present or absent depends on the kinetic equation of the reaction:

$$i = av^b \quad (1)$$

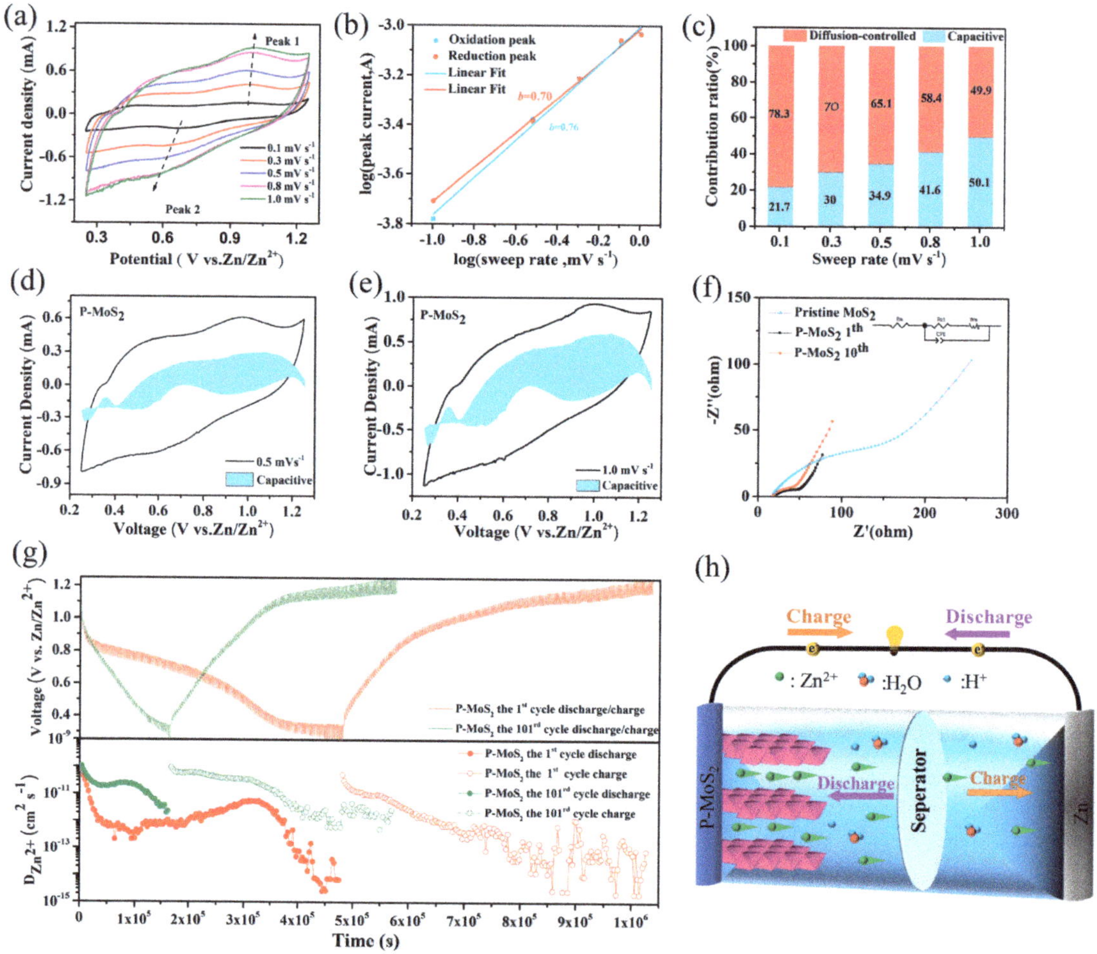

Figure 6. Electrochemical kinetics characterization for the P-MoS_2 electrode. (**a**) CV curves at various scan rates. (**b**) Log (i) versus log (v) plots of the redox peaks corresponding to the CV data. (**c**) Capacitive separation curves at 0.5 mV s^{-1} (**d**) Capacitive separation curves at 0.5 mV s^{-1}. (**e**) Capacitive separation curves at 1 mV s^{-1}. (**f**) EIS spectra, insert image is the fitted equivalent circuit models. (**g**) Discharge/charge GITT profiles and corresponding D_{Zn}. (**h**) Schematic diagram for Zn-storage mechanism of P-MoS_2.

In between, i is the peak current, the unit is A, v is the sweep speed, the unit is mV/s, a and b are the adjustment parameters, and b is the slope value of the log (i) vs. log (b) graph. In particular, b = 0.5 indicates that the discharge-specific capacity belongs to the diffusion process control, and b = 1 implies the pseudocapacitance control. When b is between 0.5–1, there is both diffusion and pseudocapacitance behavior [56]. Obviously, Figure 6b indicates the b-values of peak 1 and peak 2, which are 0.76 and 0.70, showing that the redox reaction of the P-MoS_2 electrode consists of diffusion and capacitive processes. Additionally, the capacitive contribution and diffusion contributions of P-MoS_2 may be evaluated based on the equation proposed by Dunn [57,58]:

$$i(V) = k_1 v + k_2 v^{1/2} \quad (2)$$

$$i(V)/v^{1/2} = k_1 v^{1/2} \quad (3)$$

At a certain voltage (v), i is the total current response, where $k_2 v^{1/2}$ and $k_1 v$ are diffusion-controlled and capacitance-dominated contributions. Figure 6c–e reveals the respective contribution rates at different scan rates. Notably, the capacitive contribution increases from 21.7% at 0.1 mV s^{-1} to 50.1% at 1 mV s^{-1}. The pseudocapacitance effect is slowly dominant, indicating that the surface reaction is faster than the internal diffusion reaction, which accelerates the intercalation and extraction of Zn^{2+} in the material. Moreover, the diffusion characteristics of P-MoS_2 are more obvious, which can be attributed to the marigold-shaped structure and wide layer spacing [59]. The impedance of P-MoS_2 and pristine MoS_2 is evaluated by EIS, as shown in Figure 6f. The magnitude of the charge transfer resistance is reflected in the semicircle radius of the high-frequency zone (Rct, Table S4). Clearly, results show that the charge transfer impedance of P-MoS_2 is measured at 42.82 Ω, which is lower than the pristine MoS_2 under the same conditions. In addition, Rcts of pristine MoS_2 are markedly over 180 Ω, indicating that expanded layer spacing and S vacancies improve charge-transfer dynamics. Because of the activation of the active material, the charge transfer impedance of P-MoS_2 was lowered during the first ten cycles.

To examine the solid-state diffusion kinetics of Zn^{2+} intercalation in P-MoS_2, we investigated GITT, which has been widely used to assess ionic diffusivity, offering insight into electrode kinetics. Figure 6g shows the GITT of P-MoS_2 before and after 100 cycles, respectively; obviously the P-MoS_2 cell mainly provides a low Zn^{2+} diffusion coefficient of around 10^{-15}–10^{-10} cm^2 s^{-1} at the 1st cycle. After 100 cycles, the diffusion coefficient stability was increased to 10^{-12}–10^{-10} cm^2 s^{-1}, which is attributed to the activation of the electrode and the opening of ion channels during the numerous discharging/charging progress [60,61]. The above electrochemical analysis shows that the Zn^{2+} insertion kinetics of P-MoS_2 is accelerated, leading to a higher specific capacity. When the P atoms are embedded in the MoS_2 lattice, the expanded layer spacing of MoS_2 allows for easy Zn^{2+} desertion. In addition, the P-embedding induces 1T-MoS_2 and abundant sulfur vacancies.

The material has excellent electrical conductivity and facilitates electron transfer. In addition, P-MoS_2 possesses a large amount of interlayer water and substantially increased hydrophilicity. As demonstrated in previous studies, interlayer water can act as an electrostatic shield, weakening the interaction of Zn^{2+} with the host material's framework, lowering the diffusion energy barrier, and accelerating the migration efficiency of Zn^{2+} [62,63]. Benefiting from these synergistic effects, P-MoS_2 exhibits satisfactory performance. On the basis of the aforementioned tests and analysis, the electrochemical mechanism of P-MoS_2 proposed in Figure 6h is shown, and the various electrochemical reactions that may occur with P-MoS_2 and Zn are classified as the following:

$$\text{Cathode}: xZn^{2+} + 2xe^{-} + P - MoS_2 \leftrightarrow Zn_xP - MoS_2 \quad (4)$$

$$\text{Anode}: xZn^{2+} + 2xe^{-} \leftrightarrow xZn \quad (5)$$

where x is the Zn^{2+} content per unit of P-MoS_2 in the insertion state. Figure S9 shows that the Faraday equation may calculate 0.65 Zn^{2+} per unit based on the discharge curve at 100 mA g^{-1}. Furthermore, 0.65 Zn^{2+} per unit was removed from the P-MoS_2 during the charging procedure.

4. Conclusions

In summary, marigold-shaped 1T-MoS_2 material with rich S vacancy and expanded interlayer spacing to 0.98 nm was fabricated via the oxygen-assisted method, which found the 1T phase content (53%) and extensively analyzed their performance as cathode materials for ZIBs. Meanwhile, the electrochemical investigation revealed that the S vacancy and interlayer spacing generated by P-embedding are particularly advantageous to the rapid diffusion of Zn^{2+} in P-MoS_2 and boost the Zn^{2+} storage capacity of the marigold-shaped nanosheets. In comparison to the pure MoS_2 equivalent, P-MoS_2 nanosheets have a great specific capacity and display outstanding continuous cycle capabilities in ZIBs. When the current density is 0.1 A g^{-1}, the discharge capacity can reach 249 mAh g^{-1}. In particular, it displayed a high specific discharge capacity of 105 mAh g^{-1} with a capacity retention of 70% after 3260 cycles at the current density of 0.5 A g^{-1} and an excellent capacity retention of 70 mAh g^{-1} at 1 A g^{-1}. This work offers new ideas for designing MoS_2 as cathode materials for ZIBs.

Supplementary Materials: The following supporting information can be downloaded at: https://www.mdpi.com/article/10.3390/nano13071185/s1. Figure S1. SEM images of the sample of Pristine MoS2; Figure S2. TThe corresponding atomic intensity profile along the dotted red line for P-MoS2; Figure S3. TEM images of (a,b) Pristine MoS2; Figure S4. Line scan of the HRTEM image; Figure S5. EDS spectrum; Figure S6. Contact angles with water for (a) P-MoS2 and (b) Pristine MoS2; Figure S7. XPS spectra of full scan for P-MoS2; Figure S8. The initial five CV curves of (a) Pristine MoS2 and (b) CV curves at various scan rates of Pristine MoS2; Figure S9. Inital charge-discharge profile of P-MoS2 nanosheets at 0.1A g-1; Table S1. Phase content of Mo 3d in each sample; Table S2. Atomic percentages of P-MoS2 by XPS measurement; Table S3. Comparisons of performance of MoS2 synthesized under different conditions in neutral media; Table S4. Charge transfer resistance of MoS2 samples. References [64,65] are cited in Supplementary Materials.

Author Contributions: Conceptualization, X.L. and Y.C. (Yong Cheng); methodology, X.L. and Q.X.; software, L.W.; validation, L.W. and Z.Z.; formal analysis, L.L.; investigation, Z.Z.; resources, X.L.; data curation, Y.C. (Yong Chen); writing—original draft preparation, Q.X.; writing—review and editing, X.L.; visualization, Y.C. (Yong Chen); supervision, Y.C. (Yong Cheng); project administration, L.L.; funding acquisition, X.L. All authors have read and agreed to the published version of the manuscript.

Funding: This work was supported by the National Natural Science Foundation of China (12164013, 51662004).

Data Availability Statement: Not applicable.

Conflicts of Interest: The authors declare that they have no known competing financial interests or personal relationships that could have appeared to influence the work reported in this paper.

References

1. Pan, H.; Shao, Y.; Yan, P.; Cheng, Y.; Han, K.S.; Nie, Z.; Wang, C.; Yang, J.; Li, X.; Bhattacharya, P.; et al. Reversible aqueous zinc/manganese oxide energy storage from conversion reactions. *Nat. Energy* **2016**, *1*, 16039. [CrossRef]
2. Kundu, D.; Adams, B.D.; Duffort, V.; Vajargah, S.H.; Nazar, L.F. A high-capacity and long-life aqueous rechargeable zinc battery using a metal oxide intercalation cathode. *Nat. Energy* **2016**, *1*, 16119. [CrossRef]
3. Huang, Y.; Chang, Z.; Liu, W.; Huang, W.; Dong, L.; Kang, F.; Xu, C. Layer-by-layer zinc metal anodes to achieve long-life zinc-ion batteries. *Chem. Eng. J.* **2022**, *431*, 133902. [CrossRef]
4. Zeng, X.; Hao, J.; Wang, Z.; Mao, J.; Guo, Z. Recent progress and perspectives on aqueous Zn-based rechargeable batteries with mild aqueous electrolytes. *Energy Storage Mater.* **2019**, *20*, 410–437. [CrossRef]
5. Li, H.; Yang, Q.; Mo, F.; Liang, G.; Liu, Z.; Tang, Z.; Ma, L.; Liu, J.; Shi, Z.; Zhi, C. MoS_2 nanosheets with expanded interlayer spacing for rechargeable aqueous Zn-ion batteries. *Energy Storage Mater.* **2019**, *19*, 94–101. [CrossRef]
6. Liang, H.; Cao, Z.; Ming, F.; Zhang, W.; Anjum, D.H.; Cui, Y.; Cavallo, L.; Alshareef, H.N. Aqueous Zinc-Ion Storage in MoS_2 by Tuning the Intercalation Energy. *Nano Lett.* **2019**, *19*, 3199–3206. [CrossRef]
7. Tansel, B. Significance of thermodynamic and physical characteristics on permeation of ions during membrane separation: Hydrated radius, hydration free energy and viscous effects. *Sep. Purif. Technol.* **2012**, *86*, 119–126. [CrossRef]
8. Xia, C.; Guo, J.; Li, P.; Zhang, X.; Alshareef, H.N. Highly Stable Aqueous Zinc-Ion Storage Using a Layered Calcium Vanadium Oxide Bronze Cathode. *Angew. Chem. Int. Ed.* **2018**, *57*, 3943–3948. [CrossRef]
9. Li, C.; Xie, X.; Liang, S.; Zhou, J. Issues and Future Perspective on Zinc Metal Anode for Rechargeable Aqueous Zinc-ion Batteries. *Energy Environ. Mater.* **2020**, *3*, 146–159. [CrossRef]
10. Zhang, X.; Tang, Y.; He, P.; Zhang, Z.; Chen, T. Edge-rich vertical graphene nanosheets templating V_2O_5 for highly durable zinc ion battery. *Carbon* **2021**, *172*, 207–213. [CrossRef]

11. Lee, W.S.V.; Xiong, T.; Wang, X.; Xue, J. Unraveling MoS_2 and Transition Metal Dichalcogenides as Functional Zinc-Ion Battery Cathode: A Perspective. *Small Methods* **2021**, *5*, 2000815. [CrossRef]
12. Sarkar, D.; Das, D.; Das, S.; Kumar, A.; Patil, S.; Nanda, K.K.; Sarma, D.D.; Shukla, A. Expanding Interlayer Spacing in MoS_2 for Realizing an Advanced Supercapacitor. *ACS Energy Lett.* **2019**, *4*, 1602–1609. [CrossRef]
13. Liu, X.; Zhang, G.; Zhang, Y.-W. Thermal conduction across the one-dimensional interface between a MoS_2 monolayer and metal electrode. *Nano Res.* **2016**, *9*, 2372–2383. [CrossRef]
14. Jin Jang, Y.; Park, S.-K. Rational design of hierarchical Ni-Mo bimetallic Selenide/N-doped carbon microspheres toward high–performance potassium ion batteries. *Appl. Surf. Sci.* **2022**, *583*, 152491. [CrossRef]
15. Zhao, C.; Yu, C.; Zhang, M.; Sun, Q.; Li, S.; Norouzi Banis, M.; Han, X.; Dong, Q.; Yang, J.; Wang, G.; et al. Enhanced sodium storage capability enabled by super wide-interlayer-spacing MoS_2 integrated on carbon fibers. *Nano Energy* **2017**, *41*, 66–74. [CrossRef]
16. Liu, W.; Hao, J.; Xu, C.; Mou, J.; Dong, L.; Jiang, F.; Kang, Z.; Wu, J.; Jiang, B.; Kang, F. Investigation of zinc ion storage of transition metal oxides, sulfides, and borides in zinc ion battery systems. *Chem. Commun.* **2017**, *53*, 6872–6874. [CrossRef]
17. Khan, M.F.; Miriyala, N.; Lee, J.; Hassanpourfard, M.; Kumar, A.; Thundat, T. Heat capacity measurements of sub-nanoliter volumes of liquids using bimaterial microchannel cantilevers. *Appl. Phys. Lett.* **2016**, *108*, 211906. [CrossRef]
18. Oh, H.G.; Park, S.-K. Co-MOF derived $MoSe_2$@$CoSe_2$/N-doped carbon nanorods as high-performance anode materials for potassium ion batteries. *Int. J. Energy Res.* **2022**, *46*, 10677–10688. [CrossRef]
19. Shao, F.; Huang, Y.; Wang, X.; Li, Z.; Huang, X.; Huang, W.; Dong, L.; Kang, F.; Liu, W.; Xu, C. MoS_2 with high 1T phase content enables fast reversible zinc-ion storage via pseudocapacitance. *Chem. Eng. J.* **2022**, *448*, 137688. [CrossRef]
20. Liu, J.; Gong, N.; Peng, W.; Li, Y.; Zhang, F.; Fan, X. Vertically aligned 1 T phase MoS_2 nanosheet array for high-performance rechargeable aqueous Zn-ion batteries. *Chem. Eng. J.* **2022**, *428*, 130981. [CrossRef]
21. Acerce, M.; Voiry, D.; Chhowalla, M. Metallic 1T phase MoS_2 nanosheets as supercapacitor electrode materials. *Nat. Nanotechnol.* **2015**, *10*, 313–318. [CrossRef]
22. Liu, F.; Li, L.; Xu, S.; Guo, J.; Ling, Y.; Zhang, Y.; Gong, W.; Wei, L.; Wang, C.; Zhang, Q.; et al. Cobalt-doped $MoS_2{\cdot}nH_2O$ nanosheets induced heterogeneous phases as high-rate capability and long-term cyclability cathodes for wearable zinc-ion batteries. *Energy Storage Mater.* **2023**, *55*, 1–11. [CrossRef]
23. Li, S.; Liu, Y.; Zhao, X.; Shen, Q.; Zhao, W.; Tan, Q.; Zhang, N.; Li, P.; Jiao, L.; Qu, X. Sandwich-Like Heterostructures of MoS_2/Graphene with Enlarged Interlayer Spacing and Enhanced Hydrophilicity as High-Performance Cathodes for Aqueous Zinc-Ion Batteries. *Adv. Mater.* **2021**, *33*, 2007480. [CrossRef]
24. Li, S.; Liu, Y.; Zhao, X.; Cui, K.; Shen, Q.; Li, P.; Qu, X.; Jiao, L. Molecular Engineering on MoS_2 Enables Large Interlayers and Unlocked Basal Planes for High-Performance Aqueous Zn-Ion Storage. *Angew. Chem. Int. Ed.* **2021**, *60*, 20286–20293. [CrossRef]
25. Wang, S.; Zhang, D.; Li, B.; Zhang, C.; Du, Z.; Yin, H.; Bi, X.; Yang, S. Ultrastable In-Plane 1T–2H MoS_2 Heterostructures for Enhanced Hydrogen Evolution Reaction. *Adv. Energy Mater.* **2018**, *8*, 1801345. [CrossRef]
26. Huang, X.; Leng, M.; Xiao, W.; Li, M.; Ding, J.; Tan, T.L.; Lee, W.S.V.; Xue, J. Activating Basal Planes and S-Terminated Edges of MoS_2 toward More Efficient Hydrogen Evolution. *Adv. Funct. Mater.* **2017**, *27*, 1604943. [CrossRef]
27. Nipane, A.; Karmakar, D.; Kaushik, N.; Karande, S.; Lodha, S. Few-Layer MoS_2 p-Type Devices Enabled by Selective Doping Using Low Energy Phosphorus Implantation. *ACS Nano* **2016**, *10*, 2128–2137. [CrossRef]
28. Zhou, M.; Cheng, L.; Han, B.; Zhang, H.; Chen, J.; Xie, F.; Wang, N.; Jin, Y.; Meng, H. Cobalt-doped molybdenum disulfide with rich defects and extended layered structure for rechargeable zinc-ion batteries. *J. Alloy Compd.* **2022**, *916*, 165487. [CrossRef]
29. Hu, J.; Yuan, J.; Zhao, L.; Li, G.; Chen, D.; Han, W.; Chu, Y.; Cui, X.; Li, C.; Zhang, Y. Few-layered MoS_2 with S-vacancies anchored on N-doped carbon flower for high performance sodium storage. *J. Alloy Compd.* **2022**, *895*, 162514. [CrossRef]
30. Wang, X.; Zhang, Y.; Si, H.; Zhang, Q.; Wu, J.; Gao, L.; Wei, X.; Sun, Y.; Liao, Q.; Zhang, Z.; et al. Single-Atom Vacancy Defect to Trigger High-Efficiency Hydrogen Evolution of MoS_2. *J. Am. Chem. Soc.* **2020**, *142*, 4298–4308. [CrossRef]
31. Xu, W.; Sun, C.; Zhao, K.; Cheng, X.; Rawal, S.; Xu, Y.; Wang, Y. Defect engineering activating (Boosting) zinc storage capacity of MoS_2. *Energy Storage Mater.* **2019**, *16*, 527–534. [CrossRef]
32. Xie, J.; Zhang, J.; Li, S.; Grote, F.; Zhang, X.; Zhang, H.; Wang, R.; Lei, Y.; Pan, B.; Xie, Y. Controllable Disorder Engineering in Oxygen-Incorporated MoS_2 Ultrathin Nanosheets for Efficient Hydrogen Evolution. *J. Am. Chem. Soc.* **2013**, *135*, 17881–17888. [CrossRef]
33. Xie, J.; Zhang, H.; Li, S.; Wang, R.; Sun, X.; Zhou, M.; Zhou, J.; Lou, X.W.; Xie, Y. Defect-Rich MoS_2 Ultrathin Nanosheets with Additional Active Edge Sites for Enhanced Electrocatalytic Hydrogen Evolution. *Adv. Mater.* **2013**, *25*, 5807–5813. [CrossRef]
34. Wang, D.; Pan, Z.; Wu, Z.; Wang, Z.; Liu, Z. Hydrothermal synthesis of MoS_2 nanoflowers as highly efficient hydrogen evolution reaction catalysts. *J. Power Sources* **2014**, *264*, 229–234. [CrossRef]
35. Nguyen, T.P.; Kim, I.T. W_2C/WS_2 Alloy Nanoflowers as Anode Materials for Lithium-Ion Storage. *Nanomaterials* **2020**, *10*, 1336. [CrossRef]
36. Song, H.; Li, T.; He, T.; Wang, Z.; Fang, D.; Wang, Y.; Li, X.L.; Zhang, D.; Hu, J.; Huang, S. Cooperative catalytic Mo-S-Co heterojunctions with sulfur vacancies for kinetically boosted lithium-sulfur battery. *Chem. Eng. J.* **2022**, *450*, 138115. [CrossRef]
37. Ma, D.; Li, Y.; Mi, H.; Luo, S.; Zhang, P.; Lin, Z.; Li, J.; Zhang, H. Robust SnO_{2-x} Nanoparticle-Impregnated Carbon Nanofibers with Outstanding Electrochemical Performance for Advanced Sodium-Ion Batteries. *Angew. Chem. Int. Ed.* **2018**, *57*, 8901–8905. [CrossRef]

38. Leng, K.; Chen, Z.; Zhao, X.; Tang, W.; Tian, B.; Nai, C.T.; Zhou, W.; Loh, K.P. Phase Restructuring in Transition Metal Dichalcogenides for Highly Stable Energy Storage. *ACS Nano* **2016**, *10*, 9208–9215. [CrossRef]
39. Liu, H.; Su, D.; Zhou, R.; Sun, B.; Wang, G.; Qiao, S.Z. Highly Ordered Mesoporous MoS_2 with Expanded Spacing of the (002) Crystal Plane for Ultrafast Lithium Ion Storage. *Adv. Energy Mater.* **2012**, *2*, 970–975. [CrossRef]
40. Shi, S.; Sun, Z.; Hu, Y.H. Synthesis, stabilization and applications of 2-dimensional 1T metallic MoS_2. *J. Mater. Chem. A* **2018**, *6*, 23932–23977. [CrossRef]
41. Guo, Y.; Sun, D.; Ouyang, B.; Raja, A.; Song, J.; Heinz, T.F.; Brus, L.E. Probing the Dynamics of the Metallic-to-Semiconducting Structural Phase Transformation in MoS_2 Crystals. *Nano Lett.* **2015**, *15*, 5081–5088. [CrossRef]
42. Fang, Y.; Pan, J.; He, J.; Luo, R.; Wang, D.; Che, X.; Bu, K.; Zhao, W.; Liu, P.; Mu, G.; et al. Structure Re-determination and Superconductivity Observation of Bulk 1T MoS_2. *Angew. Chem. Int. Ed.* **2018**, *57*, 1232–1235. [CrossRef]
43. Yin, Y.; Han, J.; Zhang, Y.; Zhang, X.; Xu, P.; Yuan, Q.; Samad, L.; Wang, X.; Wang, Y.; Zhang, Z.; et al. Contributions of Phase, Sulfur Vacancies, and Edges to the Hydrogen Evolution Reaction Catalytic Activity of Porous Molybdenum Disulfide Nanosheets. *J. Am. Chem. Soc.* **2016**, *138*, 7965–7972. [CrossRef]
44. Xue, H.; Meng, A.; Chen, C.; Xue, H.; Li, Z.; Wang, C. Phosphorus-doped MoS_2 with sulfur vacancy defects for enhanced electrochemical water splitting. *Sci. China Mater.* **2022**, *65*, 712–720. [CrossRef]
45. Liu, H.; Wang, J.-G.; Hua, W.; You, Z.; Hou, Z.; Yang, J.; Wei, C.; Kang, F. Boosting zinc-ion intercalation in hydrated MoS_2 nanosheets toward substantially improved performance. *Energy Storage Mater.* **2021**, *35*, 731–738. [CrossRef]
46. Angamuthu, G.; Rengarajan, V. MoS_2 mediated nitrogen enriched composite material for high and fast Li-ion storage. *Appl. Surf. Sci.* **2020**, *525*, 146437. [CrossRef]
47. Sheng, Z.; Qi, P.; Lu, Y.; Liu, G.; Chen, M.; Gan, X.; Qin, Y.; Hao, K.; Tang, Y. Nitrogen-Doped Metallic MoS_2 Derived from a Metal–Organic Framework for Aqueous Rechargeable Zinc-Ion Batteries. *ACS Appl. Mater. Interfaces* **2021**, *13*, 34495–34506. [CrossRef]
48. Eda, G.; Yamaguchi, H.; Voiry, D.; Fujita, T.; Chen, M.; Chhowalla, M. Photoluminescence from Chemically Exfoliated MoS_2. *Nano Lett.* **2011**, *11*, 5111–5116. [CrossRef]
49. Cho, J.; Ryu, S.; Gong, Y.J.; Pyo, S.; Yun, H.; Kim, H.; Lee, J.; Yoo, J.; Kim, Y.S. Nitrogen-doped MoS_2 as a catalytic sulfur host for lithium-sulfur batteries. *Chem. Eng. J.* **2022**, *439*, 135568. [CrossRef]
50. Liu, J.; Wang, Z.; Li, J.; Cao, L.; Lu, Z.; Zhu, D. Structure Engineering of MoS_2 via Simultaneous Oxygen and Phosphorus Incorporation for Improved Hydrogen Evolution. *Small* **2020**, *16*, 1905738. [CrossRef]
51. Wu, W.; Zhao, Y.; Li, S.; He, B.; Liu, H.; Zeng, X.; Zhang, J.; Wang, G. P doped MoS_2 nanoplates embedded in nitrogen doped carbon nanofibers as an efficient catalyst for hydrogen evolution reaction. *J. Colloid Interface Sci.* **2019**, *547*, 291–298. [CrossRef]
52. Chae, M.S.; Heo, J.W.; Lim, S.-C.; Hong, S.-T. Electrochemical Zinc-Ion Intercalation Properties and Crystal Structures of $ZnMo_6S_8$ and $Zn_2Mo_6S_8$ Chevrel Phases in Aqueous Electrolytes. *Inorg. Chem.* **2016**, *55*, 3294–3301. [CrossRef]
53. Cheng, Y.; Luo, L.; Zhong, L.; Chen, J.; Li, B.; Wang, W.; Mao, S.X.; Wang, C.; Sprenkle, V.L.; Li, G.; et al. Highly Reversible Zinc-Ion Intercalation into Chevrel Phase Mo_6S_8 Nanocubes and Applications for Advanced Zinc-Ion Batteries. *ACS Appl. Mater. Interfaces* **2016**, *8*, 13673–13677. [CrossRef]
54. Zhang, K.; Jin, B.; Gao, Y.; Zhang, S.; Shin, H.; Zeng, H.; Park, J.H. Aligned Heterointerface-Induced 1T-MoS_2 Monolayer with Near-Ideal Gibbs Free for Stable Hydrogen Evolution Reaction. *Small* **2019**, *15*, 1804903. [CrossRef]
55. Xu, J.; Dong, Z.; Huang, K.; Wang, L.; Wei, Z.; Yu, L.; Wu, X. Flexible design of large layer spacing V-MoS_2@C cathode for high-energy zinc-ion battery storage. *Scr. Mater.* **2022**, *209*, 114368. [CrossRef]
56. He, P.; Yan, M.; Zhang, G.; Sun, R.; Chen, L.; An, Q.; Mai, L. Layered VS_2 Nanosheet-Based Aqueous Zn Ion Battery Cathode. *Adv. Energy Mater.* **2017**, *7*, 1601920. [CrossRef]
57. Augustyn, V.; Come, J.; Lowe, M.A.; Kim, J.W.; Taberna, P.-L.; Tolbert, S.H.; Abruña, H.D.; Simon, P.; Dunn, B. High-rate electrochemical energy storage through Li^+ intercalation pseudocapacitance. *Nat. Mater.* **2013**, *12*, 518–522. [CrossRef]
58. Augustyn, V.; Simon, P.; Dunn, B. Pseudocapacitive oxide materials for high-rate electrochemical energy storage. *Energy Environ. Sci.* **2014**, *7*, 1597–1614. [CrossRef]
59. Cai, C.; Tao, Z.; Zhu, Y.; Tan, Y.; Wang, A.; Zhou, H.; Yang, Y. A nano interlayer spacing and rich defect 1T-MoS_2 as cathode for superior performance aqueous zinc-ion batteries. *Nanoscale Adv.* **2021**, *3*, 3780–3787. [CrossRef]
60. Yao, J.; Jin, T.; Li, Y.; Xiao, S.; Huang, B.; Jiang, J. Electrochemical performance of $Fe_2(SO_4)_3$ as a novel anode material for lithium-ion batteries. *J. Alloy Compd.* **2021**, *886*, 161238. [CrossRef]
61. Yao, J.; Yang, Y.; Li, Y.; Jiang, J.; Xiao, S.; Yang, J. Interconnected α-Fe_2O_3 nanoparticles prepared from leaching liquor of tin ore tailings as anode materials for lithium-ion batteries. *J. Alloy Compd.* **2021**, *855*, 157288. [CrossRef]
62. Wu, T.; Zhu, K.; Qin, C.; Huang, K. Unraveling the role of structural water in bilayer V_2O_5 during Zn^{2+}-intercalation: Insights from DFT calculations. *J. Mater. Chem. A* **2019**, *7*, 5612–5620. [CrossRef]
63. Yan, M.; He, P.; Chen, Y.; Wang, S.; Wei, Q.; Zhao, K.; Xu, X.; An, Q.; Shuang, Y.; Shao, Y.; et al. Water-Lubricated Intercalation in $V_2O_5{\cdot}nH_2O$ for High-Capacity and High-Rate Aqueous Rechargeable Zinc Batteries. *Adv. Mater.* **2018**, *30*, 1703725. [CrossRef]

54. Zhang, N.; Cheng, F.; Liu, Y.; Zhao, Q.; Lei, K.; Chen, C.; Liu, X.; Chen, J. Cation-Deficient Spinel $ZnMn_2O_4$ Cathode in $Zn(CF_3SO_3)_2$ Electrolyte for Rechargeable Aqueous Zn-Ion Battery. *J. Am. Chem. Soc.* **2016**, *138*, 12894–12901. [CrossRef]
55. Shaju, K.M.; Subba Rao, G.V.; Chowdari, B.V.R. Li ion kinetic studies on spinel cathodes, $Li(M_{1/6}Mn_{11/6})O_4$ (M = Mn, Co, CoAl) by GITT and EIS. *J. Mater. Chem.* **2003**, *13*, 106–113. [CrossRef]

nanomaterials

MDPI

Article

The Surface Properties of Implant Materials by Deposition of High-Entropy Alloys (HEAs)

Khalid Usman [1,†], Doori Kang [1,†], Geonwoo Jeong [1], Khurshed Alam [1], Athira Raveendran [1], Jinhui Ser [1], Woohyung Jang [2,*] and Hoonsung Cho [1,*]

1 School of Materials Science & Engineering, Chonnam National University, Gwangju 61186, Republic of Korea
2 Department of Prosthodontics, School of Dentistry, Chonnam National University, Gwangju 61186, Republic of Korea
* Correspondence: awj568@jnu.ac.kr (W.J.); cho.hoonsung@jnu.ac.kr (H.C.)
† These authors contributed equally to this work.

Abstract: High-entropy alloys (HEAs) contain more than five alloying elements in a composition range of 5–35% and with slight atomic size variation. Recent narrative studies on HEA thin films and their synthesis through deposition techniques such as sputtering have highlighted the need for determining the corrosion behaviors of such alloys used as biomaterials, for example, in implants. Coatings composed of biocompatible elements such as titanium, cobalt, chrome, nickel, and molybdenum at the nominal composition of $Co_{30}Cr_{20}Ni_{20}Mo_{20}Ti_{10}$ were synthesized by means of high-vacuum radiofrequency magnetron (HVRF) sputtering. In scanning electron microscopy (SEM) analysis, the coating samples deposited with higher ion densities were thicker than those deposited with lower ion densities (thin films). The X-ray diffraction (XRD) results of the thin films heat treated at higher temperatures, i.e., 600 and 800 °C, revealed a low degree of crystallinity. In thicker coatings and samples without heat treatment, the XRD peaks were amorphous. The samples coated at lower ion densities, i.e., 20 μAcm^{-2}, and not subjected to heat treatment yielded superior results in terms of corrosion and biocompatibility among all the samples. Heat treatment at higher temperatures led to alloy oxidation, thus compromising the corrosion property of the deposited coatings.

Keywords: high-entropy alloys (HEAs); sputtering; corrosion; implants; biomaterial

1. Introduction

High-entropy alloys (HEAs) are emerging advanced materials, and they are also called multicomponent alloys, multi-principal-element alloys, and compositionally complex alloys (CCAs). These alloys were first reported in 2004 by two independent research groups, namely Yeh et al. [1] and Cantor et al. [2]. The alloy reported by the latter group was composed of an equimolar mixture of Cr, Mn, Fe, Co, and Ni (i.e., Cantor alloy [3]). Owing to the synergic property of HEAs, they appear to be a better alternative to existing biomaterials such as 316 L stainless steel, titanium, and its derivative alloys such as Ti 6Al 4V, as well as conventional alloys (CAs) such as ASTM F75, F90, and F562 (HS251-Haynes Stellite), also called Vitallium [4], in terms of corrosion and biocompatibility.

Over a broad temperature range, slow diffusion, lattice distortion, and high entropy are the defining characteristics of HEAs as compared to conventional alloys (CAs). The mixture of these properties affects the corrosion resistance, tribocorrosion, strength, hardness, ductility, wear, and erosion of HEAs [5]. HEAs have the potential for application in various industries, including aerospace, energy, refractory materials, and three-dimensional (3D) printing, in addition to their use as biomaterials in implants or surgical equipment.

The constituent materials of the $Co_{30}Cr_{20}Ni_{20}Mo_{20}Ti_{10}$ HEA were selected from the perspective of biological compatibility. Co-Cr and its oxide states, e.g., Cr (VI), are considered CMR substances and cause lung cancer [6]. Ni sensitization is frequently accompanied by Co sensitization. There was no indication of Ti-specific hypersensitivity [7]. To ensure the neoformation of tissues and organs and the cells in a scaffold, the surface must be compatible anatomically, physiologically, and histologically. New bone formation can be reinforced by using bioactive signals such as genes, tethered agents, soluble factors, total growth factor (TGF-β), and its subgroups such as bone morphogenetic protein (BMP) [8,9]. Protein adhesion initiates the initial phases of corrosion in human implants. Proteins either adhere permanently to metallic implants or depart from them through flow assisted by human bodily fluids, carrying some metallic fragments in the process. The disadvantage of conventional Ti-metallic implants is that it has strength ductility mismatch with bones and also release extra low interstitial (ELI) of Ti-6Al-4V leading to phosphorus deficiency in blood and bones [10].

To synthesize the HEA, a few basic parameters must be considered, including the Hume-Rothery rule [11,12], intrinsic properties such as entropy and enthalpy of mixing [13], valence electron configuration (*VEC*) [14], and rules for single phase, crystalline, intermetallic materials [3,15,16]. HEAs can be designed by evaluating the properties of materials that constitute HEA [17,18]. Alternatively, they can be synthesized by using computing techniques such as the phase-diagram calculation [19]. The principal equations that govern the properties of HEAs are as follows:

$$\Delta S_{mix} = R\sum_{i}^{n} c_i \ ln \ c_i \tag{1}$$

$$VEC = \sum_{i=1}^{n} c_i (VEC)_i \tag{2}$$

$$\delta = 100\% \sqrt{\sum_{i=1}^{n} c_i \left(1 - \frac{r_i}{\sum_{j=1}^{n} c_j r_j}\right)^2} \tag{3}$$

$$\Delta H_{mix} = \sum_{i=1, i \neq j}^{n} \Omega_{ij} c_i c_j \tag{4}$$

$$\Omega = \frac{T_m \Delta S_{mix}}{|\Delta H_{mix}|} \tag{5}$$

where ΔS_{mix} is the entropy of mixing, R is the universal gas constant, *VEC* denotes the valence electron configuration, delta (δ) is a unitless value indicating atomic size difference, ΔH_{mix} is the enthalpy of mixing, Ω is a dimensionless parameter that denotes energy gain, T_m is the melting point, and c_i is the atomic composition. These mathematical expressions are used to evaluate the alloy systems.

In conventional methods such as casting and powder metallurgy, samples must be recast and pressed several times under vacuum to achieve homogeneity [20]. Moreover, even after these processes, the elements do not mix with complete miscibility and form amalgamations in certain microscopic regions. By contrast, sputtering, a type of physical vapor deposition (PVD) method, involves more spontaneous mixing of the constituent materials. Moreover, it does not require additional recasting and pressing steps. With sputtering, the layer thickness of HEA, which impacts the corrosion resistance and biocompatibility of the underlying substrate, can be controlled easily. In this work, corrosion analyses of sputtered HEA thick and thin films in simulated body fluid (SBF) are performed, whereas in existing studies, the corrosion properties of HEA were studied in NaCl, H_2SO_4, and seawater. During sputtering, the films are subjected to a metastable-phase energy barrier of E~0.25 eV before their final adhesion to the substrate surface, and this metastable phase is the transition region between the gasification and substrate phases [21].

2. Materials and Methods

The top-down configuration of an Artec system magnetron sputtering machine was used for HEA deposition. This machine is equipped with a target holder measuring 4.25 in. The target holder was upgraded to accommodate 20 pieces. In total, six cobalt, two titanium, and four nickel, chromium, and molybdenum were cut by means of electron discharge machining (EDM). Approximately 99.99% pure elemental discs were procured from THIFINE (the film and fine materials ©). To avoid charge accumulation at the target, the radiofrequency (RF) sputtering process was used, as illustrated in Figure 1a. The frequency ranged from 5 to 30 MHz, and the typical frequency was 13.56 MHz; electron densities ranged from 10^9 to 10^{11} cm^{-3}, and discharge pressure ranged from 0.5 to 10 mTorr [22]. Rare-earth magnets were used in the RF magnetron sputtering process. Figure 1b depicts how the magnetic field triggered a spiral momentum of the electrons around the target. This momentum increases the probability of sustained plasma, even at low pressures [23]. A 5×10^{-2} mbar vacuum was created before turning on the main valve (M/V). The samples were prepared at two different ion current densities of 2×10^{-5} A/cm^2 and 5×10^{-5} A/cm^2 [24]. Lower ion current densities reduce the likelihood of substrate oxidation [25]. After the thin HEA films were deposited by means of sputtering, the samples were post-treated in a multifunctional vacuum furnace (Ajeon Heating Industrial Co., Ltd., Namyangju-si, Republic of Korea) for 1 h at three different temperatures of 400 °C, 600 °C, and 800 °C. In this treatment, the maximum temperature was attained with a low heating curve rate, i.e., 5 °C per minute, to avoid large temperature gradients that result in tensile stresses and lead to the Kirkendall effect in coatings [26]. The treated samples were then cooled in a furnace under a 0.14 SCCM argon flow.

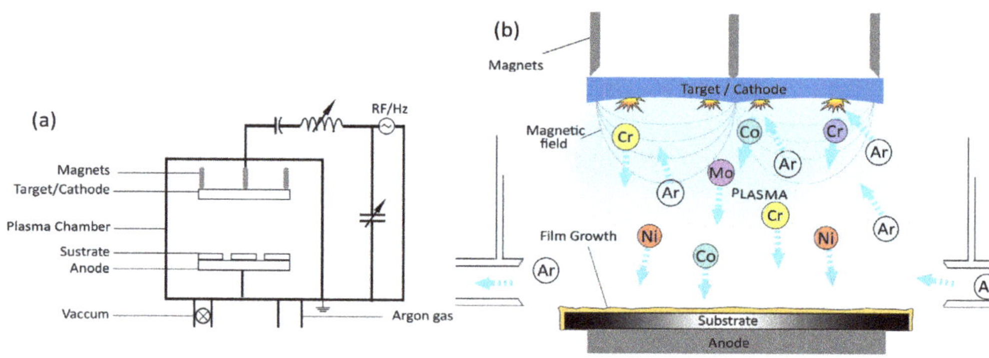

Figure 1. (**a**) Line diagram of radiofrequency (RF) magnetron sputtering; (**b**) Inside view of HEA sputtering process in the chamber.

Thermodynamic calculations were performed using mathematical expressions. From Equation (3), the average atomic radius r_{avg} of the prepared HEA $Co_{30}Cr_{20}Ni_{20}Mo_{20}Ti_{10}$ was calculated as 1.2931 Å. The atomic size difference was 9.4%. By applying Equation (2) to the XPS composition analysis result of the sample (Table 1), the nominal valence electron configuration (*VEC*) was calculated as 7.15. The enthalpy of mixing was calculated using Equation (4) (ΔH_{mix} was taken as −4.07 KJ/mol) [27,28].

Corrosion experiments were conducted using BioLogic's VSP 300 potentiostat and EC-Lab software (v11.33). For corrosion testing, the Metek-designed flat cell K0235 was used, which exposes 1 cm^2 of the working electrode surface for greater accuracy. The cell was designed according to ASTM G5 with constant argon purging [29]. The fact that the oxygen in electrolytes reaches the surface of HEA affects the rate of oxidation on the HEA. By combining Fick's law for diffusion and Faraday's law, one can calculate the cathodic current associated with oxygen diffusion [30]. A simulated body fluid called Ringer's solution was selected as the electrolyte for corrosion testing, and Bode plots were constructed to analyze the electrochemical impedance of the synthesized HEA [31–34].

Cellular metabolic activities were monitored by developing a 3-(4,5-dimethylthiazol-2-yl)-2,5-diphenyl-2H-tetrazolium bromide (MTT) assay of osteoblast cell type (3T3 cells). MC 3T3 (osteoblast cells) were taken from a cryogenic freezer and grown in an incubator for one month in regular alpha-minimum essential medium (MEM) without L-ascorbic acid and phenol red at 37 °C and with 5% CO_2 supply. After the cells grew to 80% confluence in a T-75 flask, they were sufficiently populated. Twice each week, the medium was changed, and the MTT assay was developed when the cell confluence exceeded 1 × 10^5 cells/mL [35].

Table 1. Elemental composition and orbital configuration of deposited $Co_{30}Cr_{20}Ni_{20}Mo_{20}Ti_{10}$ HEA.

Element	Atomic Percentage (%)	Binding Energy (eV)	FWHM (ΔeV)	Chi-Square (×10^6)
Cobalt	31.89	777~794		
Co $2p_{3/2}$	80.41 of 31.89	778.42	1.10	3.25
Co $2p_{1/2}$	19.59 of 31.89	793.29	1.52	
Chromium	17.42	573~585		
Cr $2p_{3/2}$	74.20 of 17.42	574.46	1.77	5.74
Cr $2p_{1/2}$	25.80 of 17.42	583.58	1.75	
Molybdenum	21.93	227~231		
Mo $3d_{5/2}$	62.60 of 21.93	227.59	0.79	3.65
Mo $3d_{3/2}$	37.40 of 21.93	229.46	1.16	
Nickel	20.51	869~852		
Ni $2p_{3/2}$	62.24 of 20.51	852.70	1.29	4.08
Ni $2p_{3/2}$	7.29 of 20.51	859.60	2.46	
Ni $2p_{1/2}$	30.47 of 20.51	871.69	2.01	
Titanium	8.24	455~462		
Ti $2p_{3/2}$	59.52 of 8.24	455.02	1.50	0.34
Ti $2p_{1/2}$	40.48 of 8.24	461.67	1.83	

Thin-Film Characterization

X-ray diffraction analysis (XRD, using Cu-K radiation) was performed to examine the crystallinity and phase structure of the synthesized HEAs. A low scan rate (1°/min) was employed to obtain a higher resolution. The step size, 2θ, and omega range of the goniometer were 0.02, 5–90°, respectively, and a voltage of 45 kV was used. By means of field-emission scanning electron microscopy, the thicknesses of the deposited thin films were measured (Model: Gemini 500). By using an Al Kα X-ray analyzer in conjunction with XPS (Model: K-ALPHA+) with a spot size of 400 µm, excitation energy of 4.36 keV, energy step of 0.05 eV, and pass energy of 100 eV, the surface chemistries of the produced HEAs were determined. Argon etching was performed before characterization, and the C1 s peak was set to 284.6 eV.

3. Results and Discussion

After sputtering with varying ion densities, SEM was used to determine the coating thickness, as shown in Figure 2. Higher ion density (5 × 10^{-5} A/cm^2) had more argon ions inside the sputtering chamber, resulting in thicker samples (the samples synthesized by this parameter are called thick hereinafter), and samples prepared at low ion density, i.e., 2 × 10^{-5} A/cm^2, had less thickness with a difference of approximately 250 nm w.r.t thick coating, labeled as thin hereinafter.

For XRD characterization, the amorphous peak of the thin films was absorbed because, according to Equation (3), the atomic difference of $Co_{30}Cr_{20}Ni_{20}Mo_{20}Ti_{10}$ was 5.57%, which is in the amorphous region for HEAs, as depicted in Figure 3 [36].

The heat treatment temperature range was 300–1000 °C for the HEA without copper. In this temperature range, the compositional difference between the dendritic ends and the center was eliminated, except in the case of the HEAs with copper in them because, in this temperature range, copper tended to segregate [37]. So, the prepared HEAs were heat treated within given temperature ranges.

The lattice distortion effect of the HEA with varying atomic sizes led to a significant loss of crystallization, which simplified its XRD patterns at high temperatures and atypically lowered the peak heights of the (310) and (200) planes at 600 °C and 800 °C, respectively [36].

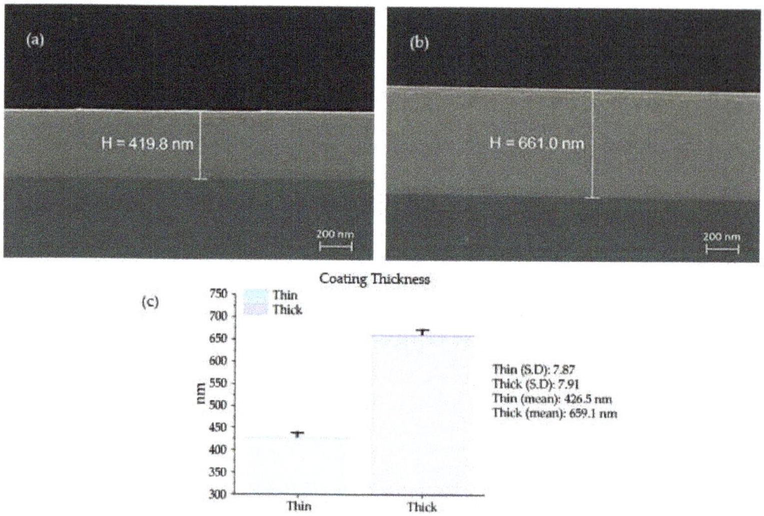

Figure 2. SEM analysis of HEA thin films synthesized at two different ion densities of (**a**) 20 μAcm^{-2} (thin film) and (**b**) 50 μAcm^{-2} (thick film). (**c**) Statistical differences between these two coatings.

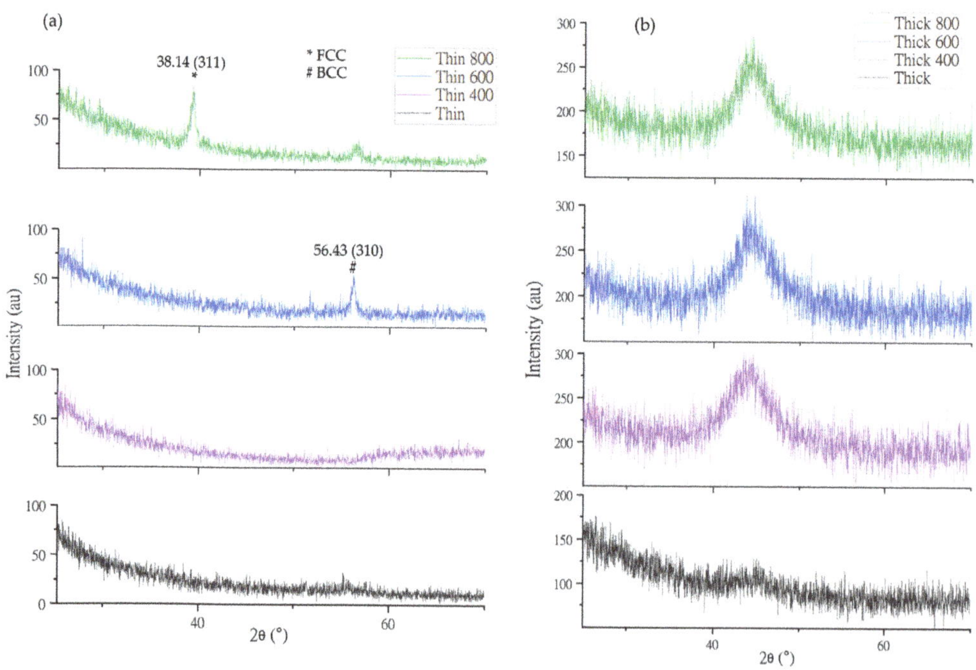

Figure 3. Amorphous XRD peaks of (**a**) thin and (**b**) thick HEA films without heat treatment (WHT) and heat treated at 400 °C, 600 °C, and 800 °C, respectively.

The XPS results of the alloy are depicted in Figure 4, along with the corresponding Gaussian peak fitting results. The XPS results were obtained using a sample that was not heat treated (Table 2) [38–42]. Figure 4a depicts the XPS survey spectrum, which confirms the coexistence of all deposited elements in the $Co_{30}Cr_{20}Ni_{20}Mo_{20}Ti_{10}$ HEA. By means of Ar etching for 17 min before the XPS examination, we decreased the heights of the O1s and C1s peaks considerably. Figure 4b depicts the XPS peaks of Cr at 574 eV and 583.5 eV, which correspond to Cr $2p_{3/2}$ and Cr $2p_{1/2}$, respectively [41,43]. These peaks were further deconvoluted into two binding energies. In Figure 4c, cobalt has two prominent peaks at 778 and 793 eV, which correlate to $Co2p_{3/2}$ and $Co2p_{1/2}$, respectively. These intensities are further split into two regions in each peak [39,44]. The peak at 230.8 eV corresponds to ($Mo3d_{3/2}$) Mo^{+4} [45] in Figure 4d, and in the same proximity, the peak centered at 227.5 eV corresponds to $Mo3d_{5/2}$ [38,40].

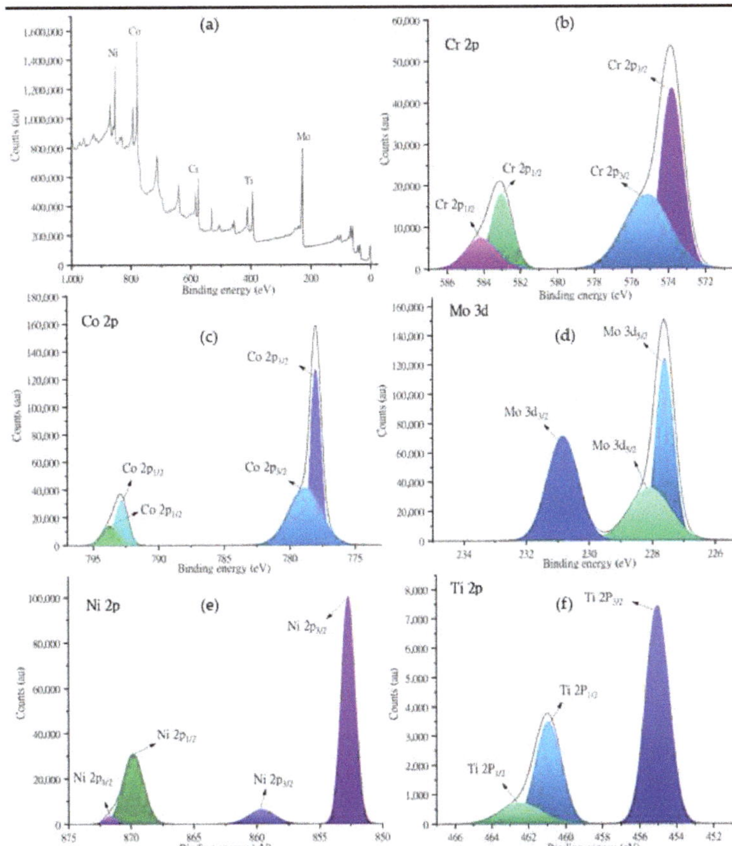

Figure 4. (**a**) XPS analysis of Co30Cr20Ni20Mo20Ti10 HEA-synthesized through magnetron sputtering and the corresponding peak fitting of (**b**) chromium, (**c**) cobalt, (**d**) molybdenum, (**e**) nickel, and (**f**) titanium.

The nickel peaks in Figure 4e were discretized into $Ni2p_{3/2}$ peaks with spin-orbital characteristics at 852 and 859 eV and an $Ni2P_{1/2}$ peak centered at 870 eV [40,44]. Similarly in Figure 4f, the Ti peaks are split into a $Ti2P_{3/2}$ peak centered at 455 eV and two $Ti2P_{1/2}$ peaks centered at 461 eV [42].

The results of the XPS studies confirmed that Mo^{6+} (Mo $3d_{5/2}$) and Mo^{4+} (Mo $3d_{3/2}$) were present in the synthesized coatings, and these ions were an integral part of the passive Cr_2O_3 films, in which Mo $3d_{5/2}$ controlled the film stability. MoO_4^{2-} and MoO_2 in these films, which had binding energies of around 227 eV and 230 eV, respectively, as listed in Table 1, acted as healers of any defects in the passive films [34].

Table 2. Electrochemical properties of HEA with reference to uncoated titanium CP-II.

Property	Uncoated	Thin Coat	Thin Coat 400	Thin Coat 600	Thin Coat 800	Thick Coat	Thick Coat 400	Thick Coat 600	Thick Coat 800
E_{corr} (mV)	−363.08	−347.46	527.04	522.49	−318.95	−698.23	163.09	−297.67	−463.64
I_{corr} (μA)	0.313	0.040	0.111	0.122	0.108	1.851	0.145	0.126	0.088
Corrosion rate (mmpy)	0.0108	0.0014	0.0038	0.0042	0.0037	0.0641	0.0050	0.0043	0.0030
Resistance (Ω)	1.16×10^6	8.69×10^6	4.74×10^6	4.28×10^6	2.96×10^6	3.77×10^5	1.12×10^6	2.36×10^6	5.26×10^6
E_{pit} (V)	–	0.569			0.575	0.629	0.434		0.486
Log I_{pass} (mA)	–	0.010			0.033	0.029	0.080		0.433

For corrosion analysis, the open-circuit potential was stabilized for 3 h, and polarization curves were drawn between −0.5 V and 1.5 V at a scan rate of 10 mV/min on titanium (CP-II) substrate. The reference electrode for the electrochemical process was Ag/AgCl with 3M NaCl/saturated AgCl filling solution, and the electrode potential of the Ag/AgCl electrode is 0.02 V against the standard calomel electrode (SCE) [46]. The electrochemical impedance spectroscopy (EIS) experiment was run before drawing the Tafel plots in Figures 5 and 6 because EIS is a non-destructive analysis. Moreover, the electrochemical properties of the synthesized HEA are summarized in Table 2.

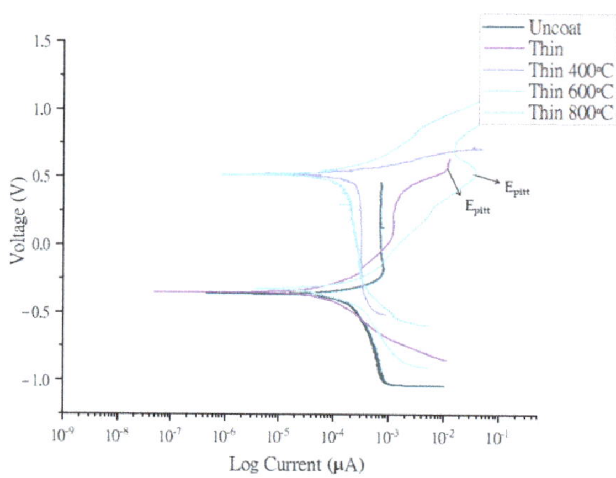

Figure 5. Polarization curves of Co30Cr20Ni20Mo20Ti10 alloy versus Ag/AgCl electrode in Hank's solution. Working electrode was deposited at 2 × 105 ion density (thin coat) without heat treatment and with heat treatment at 400 °C, 600 °C, and 800 °C.

As a biomaterial, the synthesized HEA surface coating must be characterized electrochemically to understand its behavior on implants from the perspective of avoiding inflammation through the release of ions into the human body. This is because the synergic effect of HEA and its corrosion characteristics vary significantly from those of known electrochemical processes.

The pitting corrosion on the coating surface is considered a major disadvantage for biomedical implants. In the discussion of the corrosion, we can add the fact that in the trans-passive region of the potentiodynamic curve, potential considerably greater than the corrosion potential pitting occurs (ref: ASTM-F746-0), as illustrated in Figures 5 and 6.

The uncoated sample was a substrate of commercially pure titanium grade II substrate. The polarization curves represented in Figures 5 and 6 were measured against the open-circuit potential. Ag/AgCl, and a Pt mesh were used as a reference and counter electrodes, respectively. Table 2 lists the current densities, corrosion resistance, and other important values obtained from the Tafel plots after fitting in EC-Lab software.

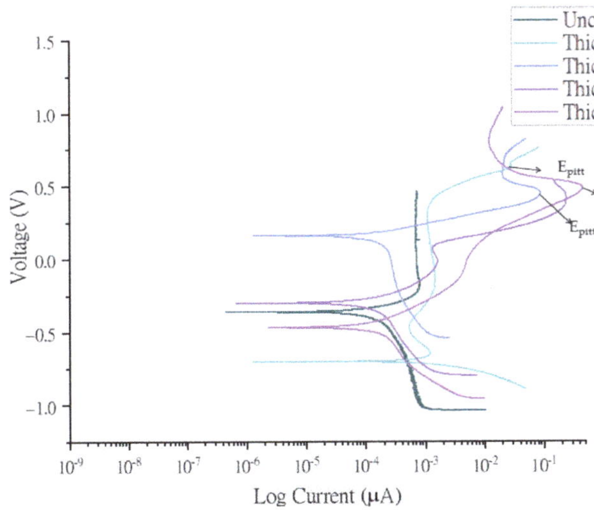

Figure 6. Polarization curves of Co30Cr20Ni20Mo20Ti10 alloy versus Ag/AgCl electrode in Hank's solution. Working electrode was deposited at 5 × 105 ion density (thin coat) without heat treatment and with heat treatment at 400 °C, 600 °C, and 800 °C.

The atomic arrangement of synthesized HEA was amorphous, and high temperature leads to thermally active movement of atoms resulting in the formation of a small-ordered crystal structure, as depicted in Figure 3. Due to this movement, the compactibility of the coating is compromised, resulting in high current densities at elevated temperatures [47], as shown in Table 2. In the synthesized thin film, Mo-O and Cr-O generate a passivation mechanism by providing a healing effect to the coating. Therefore the amounts of molybdenum and chromium in the coating gives a passivation mechanism by the Cr-rich sigma (σ) phase [16]. This is corroborated by the high corrosion resistance of $8.69 \times 10^9\ \Omega$ of the thin film [34]. Cr (VI) is released from an alloy. It is only present for a very limited time because it is quickly reduced to the trivalent state in vivo [48].

Electrochemical impedance is a measurement of the resistance of coating for a given frequency and phase of alternating current. The higher the potential of a material, the lower its electron density at that potential, which means that there is more space for electrons to flow, and the surface is therefore resistant to oxidation. Electrochemical impedance characterization was performed by analyzing the Bode plots of the samples, as depicted in Figure 7. The frequency range used for the EIS measurements was 10 kHz to 1 mHz, and 1 sinusoidal wave was composed of 10 nodes of data points. Constant phase elements (CPEs) were determined using the following equation.

$$Z_{CPE}(\omega) = \omega \frac{1}{(i\omega)^{\alpha} Q} \tag{6}$$

where ω is the angular frequency $0 \leq \alpha \leq 1$. At higher and lower frequencies, the circuit acted as capacitive and resistive, respectively.

The Bode plots show the variation of the impedance modulus and the phase angle of the CPEs in the film at the applied frequency of the sinusoidal signal. HEAs have higher impedance owing to the presence of metals that generate more ions in the resistive region, that is, at low frequencies ranging from 10^{-2} to 10^{-1}, the thick and thin 400 °C samples show less contribution of charge transfer because their magnitudes are greater than 10^2 to $10^3\ \Omega$ on a logarithmic scale.

MTT dye was analyzed using a wavelength of 570 nm to perform a cell viability analysis. Osteoblast 3T3 cells were grown on processed $Co_{30}Cr_{20}Ni_{20}Mo_{20}Ti_{10}$ HEAs, and their compatibility is shown in Figure 8. MTT analysis was performed at a significance level of 0.05 and a mean square value of 0.16. The sample heat treated at 600 °C indicated a significant difference in variability (S = 0.05). This HEA sample contained 10% Ti in three structural forms, namely alpha, beta, and alpha-beta phases, depending on the heat treatment condition and alloying additions. At 600 °C, the Ti was mostly α-Ti, leading to reduced cell viability [49]. Mo and Cr also functioned as beta stabilizers, for instance, in ASTM F 2066. Moreover, the biocompatibility of the samples decreased as their heat treatment temperature increased, as the electrical current started to flow within the cells [50]. The samples heat treated at 800 °C exhibited little crystallinity in the XRD (Figure 3), and accordingly, their biocompatibility was poor.

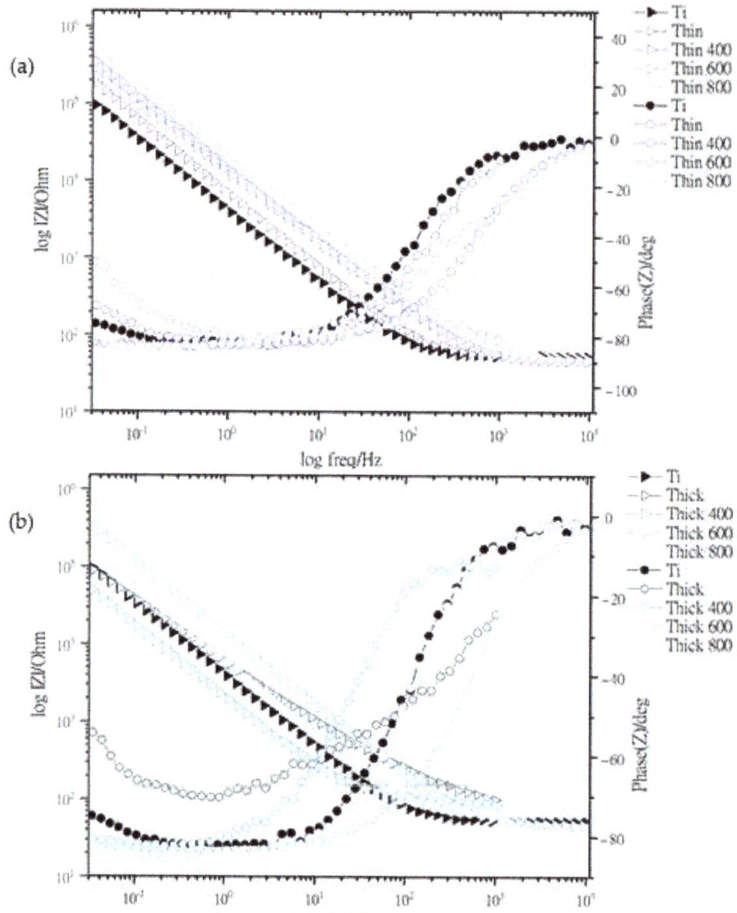

Figure 7. Impedance plots (Bode curve) of $Co_{30}Cr_{20}Ni_{20}Mo_{20}Ti_{10}$ alloy versus Ag/AgCl electrode in Hank's solution. Working electrode was deposited at (**a**) 2×10^5 ion density (thin coat) without heat treatment and with heat treatment at 400 °C, 600 °C, and 800 °C (**b**); (**a**) 5×10^5 ion density (thin coat) without heat treatment and with heat treatment at 400 °C, 600 °C, and 800 °C.

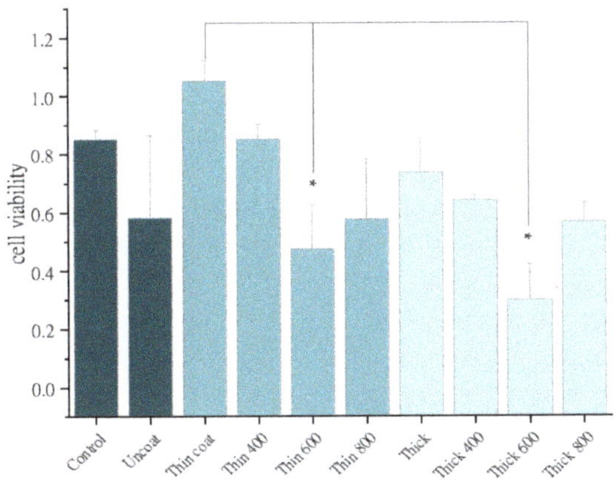

Figure 8. MTT analysis of uncoated titanium surface, coated $Co_{30}Cr_{20}Ni_{20}Mo_{20}$, and heat treated surfaces. The significant difference between thin coat and samples heat treated at 600 °C is marked with hysteric.

4. Conclusions

In this work, the thin HEA coating of 419 nm produced by sputtering at 2×10^{-5} A/cm^2 ion density and without heat treatment outperformed all the other samples in terms of corrosion resistance of 8.69×10^6 Ω-cm^2 (Table 2) and biological cell viability (Figure 8). In our analysis, we studied an HEA by considering two variables, namely coating thickness in nanoscale and heat treatment temperature. In our analysis of the EIS property of the deposited HEA thin films, the WHT thin film formed a resistive circuit that protected biological cells and allowed current flow in the extracellular region. Therefore, this coating must have had a compacted passive layer. If the chromium content of the HEA exceeded 12%, molybdenum eliminated hydroxide ions from chromium. Similarly, nickel and cobalt enhanced the strength of the passivation layer, which increased its resistance to pitting corrosion [51], as shown in Table 2 [34]. The samples coated with $Co_{30}Cr_{20}Ni_{20}Mo_{20}Ti_{10}$ at low ion density and without heat treatment produced relatively good osteoblast cell viability and corrosion resistance, as illustrated in Figure 9. This finding is solely attributable to the fact that coated samples tend to have less structure distortion than the other samples, and atomic mobility is noticeable in the XRD results from the remaining samples.

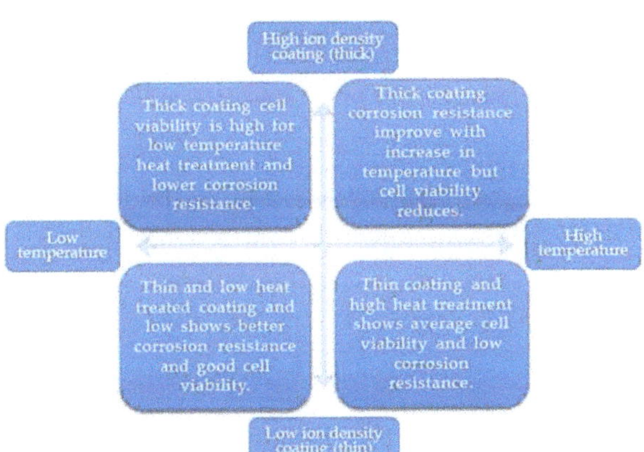

Figure 9. Effect of heat treatment and coating thickness on electrochemical property and osteoblastic cell viability.

Author Contributions: Conceptualization, K.A. and H.C.; Methodology, K.U. and G.J.; Software, D.K.; Validation, G.J., K.A., A.R. and J.S.; Formal analysis, K.U., K.A. and J.S.; Investigation, D.K. and A.R.; Resources, A.R.; Data curation, G.J.; Writing—original draft, K.U. and D.K.; Writing—review & editing, J.S. and W.J.; Visualization, K.U.; Supervision, H.C.; Project administration, W.J. and H.C.; Funding acquisition, W.J. and H.C. All authors have read and agreed to the published version of the manuscript.

Funding: This work was supported by a National Research Foundation of Korea (NRF) grant funded by the Korea Government (MSIT) (No. 2021R1F1A1064062) and by the Basic Science Research Program through the National Research Foundation of Korea (NRF) funded by the Ministry of Education (No. 2021R1I1A1A01058691).

Institutional Review Board Statement: Not applicable.

Informed Consent Statement: Not applicable.

Data Availability Statement: Data will be available upon request.

Conflicts of Interest: This research article has no conflict of interest. The authors declare that they have no conflicts of interest related to the content of this work.

References

1. Yeh, J.-W.; Chen, S.K.; Lin, S.-J.; Gan, J.-Y.; Chin, T.-S.; Shun, T.-T.; Tsau, C.-H.; Chang, S.-Y. Nanostructured High-Entropy Alloys with Multiple Principal Elements: Novel Alloy Design Concepts and Outcomes. *Adv. Eng. Mater.* **2004**, *6*, 299–303. [CrossRef]
2. Cantor, B.; Chang, I.T.H.; Knight, P.; Vincent, A.J.B. Microstructural development in equiatomic multicomponent alloys. *Mater. Sci. Eng. A* **2004**, *375*, 213–218. [CrossRef]
3. George, E.P.; Raabe, D.; Ritchie, R.O. High-Entropy Alloys. *Nat. Rev. Mater.* **2019**, *4*, 515–534. [CrossRef]
4. Geib, F.W. Vitallium Skull Plates. *J. Am. Med. Assoc.* **1941**, *117*, 8–12. [CrossRef]
5. Alvi, S.; Jarzabek, D.M.; Kohan, M.G.; Hedman, D.; Jenczyk, P.; Natile, M.M.; Vomiero, A.; Akhtar, F. Synthesis and Mechanical Characterization of a CuMoTaWV High-Entropy Film by Magnetron Sputtering. *ACS Appl. Mater. Interfaces* **2020**, *12*, 21070–21079. [CrossRef]
6. Vaicelyte, A.; Janssen, C.; Le Borgne, M.; Grosgogeat, B. Cobalt–Chromium Dental Alloys: Metal Exposures, Toxicological Risks, CMR Classification, and EU Regulatory Framework. *Crystals* **2020**, *10*, 1151. [CrossRef]
7. Linauskienė, K.; Malinauskienė, L.; Blažienė, A. Metals Are Important Contact Sensitizers: An Experience from Lithuania. *BioMed Res. Int.* **2017**, *2017*, 3964045. [CrossRef]
8. Ring, P.A. Complete Replacement Arthroplasty of the Hip by the Ring Prosthesis. *J. Bone Jt. Surg. Br. Vol.* **1968**, *50*, 720–731. [CrossRef]
9. Paralkar, V.M.; Grasser, W.A.; Riccardi, K.A.; Thompson, D.D.; Vukicevic, S. *Bone Healing Bone Morphogenetic Proteins and Beyond, in Bone Morphogenetic Proteins: Regeneration of Bone and Beyond*; Vukicevic, S., Sampath, K.T., Eds.; Birkhäuser Basel: Basel, Switzerland, 2004; pp. 293–305.
10. Gurappa, I. Characterization of different materials for corrosion resistance under simulated body fluid conditions. *Mater. Charact.* **2002**, *49*, 73–79. [CrossRef]
11. Mizutani, U. *Hume-Rothery Rules for Structurally Complex Alloy Phases*; Tylor & Francis Group: Boca Raton, FL, USA, 2010; p. 356.
12. Pei, Z.; Yin, J.; Hawk, J.A.; Alman, D.E.; Gao, M.C. Machine-learning informed prediction of high-entropy solid solution formation: Beyond the Hume-Rothery rules. *npj Comput. Mater.* **2020**, *6*, 50. [CrossRef]
13. Otto, F.; Yang, Y.; Bei, H.; George, E. Relative effects of enthalpy and entropy on the phase stability of equiatomic high-entropy alloys. *Acta Mater.* **2013**, *61*, 2628–2638. [CrossRef]
14. Chen, R.; Qin, G.; Zheng, H.; Wang, L.; Su, Y.; Chiu, Y.; Ding, H.; Guo, J.; Fu, H. Composition design of high entropy alloys using the valence electron concentration to balance strength and ductility. *Acta Mater.* **2018**, *144*, 129–137. [CrossRef]
15. Tsai, M.-H.; Yeh, J.-W. High-Entropy Alloys: A Critical Review. *Mater. Res. Lett.* **2014**, *2*, 107–123. [CrossRef]
16. Qiu, Y.; Thomas, S.; Gibson, M.A.; Fraser, H.L.; Birbilis, N. Corrosion of high entropy alloys. *npj Mater. Degrad.* **2017**, *1*, 15. [CrossRef]
17. Ye, Y.; Wang, Q.; Lu, J.; Liu, C.; Yang, Y. High-entropy alloy: Challenges and prospects. *Mater. Today* **2016**, *19*, 349–362. [CrossRef]
18. Pei, Z.; Yin, J.; Liaw, P.K.; Raabe, D. Toward the design of ultrahigh-entropy alloys via mining six million texts. *Nat. Commun.* **2023**, *14*, 54. [CrossRef]
19. George, E.P.; Curtin, W.A.; Tasan, C.C. High entropy alloys: A focused review of mechanical properties and deformation mechanisms. *Acta Mater.* **2020**, *188*, 435–474. [CrossRef]
20. Song, H.; Lee, S.; Lee, K. Thermodynamic parameters, microstructure, and electrochemical properties of equiatomic TiMoVWCr and TiMoVNbZr high-entropy alloys prepared by vacuum arc remelting. *Int. J. Refract. Met. Hard Mater.* **2021**, *99*, 105595. [CrossRef]
21. Ohring, M. Chapter 7—Substrate Surfaces and Thin-Film Nucleation. In *Materials Science of Thin Films*, 2nd ed.; Ohring, M., Ed.; Academic Press: San Diego, CA, USA, 2002; pp. 357–415.

22. Mattox, D.M. Chapter 7—Physical Sputtering and Sputter Deposition (Sputtering). In *Handbook of Physical Vapor Deposition (PVD) Processing*, 2nd ed.; Mattox, D.M., Ed.; William Andrew Publishing: Boston, MA, USA, 2010; pp. 237–286.
23. Ohring, M. Chapter 5—Plasma and Ion Beam Processing of Thin Films. In *Materials Science of Thin Films*, 2nd ed.; Ohring, M., Ed.; Academic Press: San Diego, CA, USA, 2002; pp. 203–275.
24. Morris, A.; Langari, R. *Measurement and Instrumentation: Theory and Application*; Butterworth-Heinemann: Oxford, UK, 2012.
25. Xing, Q.; Wang, H.; Chen, M.; Chen, Z.; Li, R.; Jin, P.; Zhang, Y. Mechanical Properties and Corrosion Resistance of NbTiAlSiZrNx High-Entropy Films Prepared by RF Magnetron Sputtering. *Entropy* **2019**, *21*, 396. [CrossRef]
26. Song, B.; Yang, Y.; Rabbani, M.; Yang, T.T.; He, K.; Hu, X.; Yuan, Y.; Ghildiyal, P.; Dravid, V.P.; Zachariah, M.R.; et al. In Situ Oxidation Studies of High-Entropy Alloy Nanoparticles. *ACS Nano* **2020**, *14*, 15131–15143. [CrossRef]
27. Guo, S.; Liu, C.T. Phase stability in high entropy alloys: Formation of solid-solution phase or amorphous phase. *Prog. Nat. Sci. Mater. Int.* **2011**, *21*, 433–446. [CrossRef]
28. Takeuchi, A.; Inoue, A. Classification of Bulk Metallic Glasses by Atomic Size Difference, Heat of Mixing and Period of Constituent Elements and Its Application to Characterization of the Main Alloying Element. *Mater. Trans.* **2005**, *46*, 2817–2829. [CrossRef]
29. Kinetics: Polarization and Corrosion Rates. In *Corrosion and Corrosion Control*; John Wiley & Sons: Hoboken, NJ, USA, 2008; pp. 53–82.
30. Scheiner, S.; Hellmich, C. Finite Volume model for diffusion- and activation-controlled pitting corrosion of stainless steel. *Comput. Methods Appl. Mech. Eng.* **2009**, *198*, 2898–2910. [CrossRef]
31. Manam, N.S.; Harun, W.S.W.; Shri, D.N.A.; Ghani, S.A.C.; Kurniawan, T.; Ismail, M.H.; Ibrahim, M.H.I. Study of corrosion in biocompatible metals for implants: A review. *J. Alloys Compd.* **2017**, *701*, 698–715. [CrossRef]
32. Vidal, C.V.; Muñoz, A.I. Electrochemical characterisation of biomedical alloys for surgical implants in simulated body fluids. *Corros. Sci.* **2008**, *50*, 1954–1961. [CrossRef]
33. Metikoš-Huković, M.; Katić, J.; Grubač, Z.; Škugor Rončević, I. Electrochemistry of CoCrMo Implant in Hanks' Solution and Mott-Schottky Probe of Alloy's Passive Films. *Corrosion* **2017**, *73*, 1401–1412. [CrossRef]
34. Karri, M.; Verma, A.; Singh, J.; Bonagani, S.K.; Goutam, U. Role of Chromium in Anomalous Behavior of the Passive Layer in Ni-Cr-Mo Alloys in 1 M HCl Solution. *Corrosion* **2022**, *78*, 228–238. [CrossRef]
35. Zebisch, K.; Voigt, V.; Wabitsch, M.; Brandsch, M. Protocol for effective differentiation of 3T3-L1 cells to adipocytes. *Anal. Biochem.* **2012**, *425*, 88–90. [CrossRef]
36. Yeh, J.W.; Chang, S.Y.; Hong, Y.D.; Chen, S.K.; Lin, S.J. Anomalous decrease in X-ray diffraction intensities of Cu–Ni–Al–Co–Cr–Fe–Si alloy systems with multi-principal elements. *Mater. Chem. Phys.* **2007**, *103*, 41–46. [CrossRef]
37. Li, Y.J.; Savan, A.; Kostka, A.; Stein, H.S.; Ludwig, A. Accelerated atomic-scale exploration of phase evolution in compositionally complex materials. *Mater. Horiz.* **2018**, *5*, 86–92. [CrossRef]
38. Kang, N.; Paudel, H.P.; Leuenberger, M.N.; Tetard, L.; Khondaker, S.I. Photoluminescence Quenching in Single-Layer MoS_2 via Oxygen Plasma Treatment. *J. Phys. Chem. C* **2014**, *118*, 21258–21263. [CrossRef]
39. Jacobs, G.; Chaney, J.A.; Patterson, P.M.; Das, T.K.; Davis, B.H. Fischer–Tropsch synthesis: Study of the promotion of Re on the reduction property of Co/Al_2O_3 catalysts by in situ EXAFS/XANES of Co K and Re LIII edges and XPS. *Appl. Catal. A Gen.* **2004**, *264*, 203–212. [CrossRef]
40. Hong, S.T.; Park, D.R.; Yoo, S.-J.; Kim, J.-D.; Park, H.S. Characterization of the active phase of $NiMo/Al_2O_3$ hydrodesulfurization catalysts. *Res. Chem. Intermed.* **2006**, *32*, 857–870. [CrossRef]
41. Harrison, P.G.; Lloyd, N.C.; Daniell, W. The Nature of the Chromium Species Formed during the Thermal Activation of Chromium-Promoted Tin(IV) Oxide Catalysts: An EPR and XPS Study. *J. Phys. Chem. B* **1998**, *102*, 10672–10679. [CrossRef]
42. Atuchin, V.V.; Kesler, V.G.; Pervukhina, N.V.; Zhang, Z. Ti 2p and O 1s core levels and chemical bonding in titanium-bearing oxides. *J. Electron Spectrosc. Relat. Phenom.* **2006**, *152*, 18–24. [CrossRef]
43. Bumajdad, A.; Al-Ghareeb, S.; Madkour, M.; Sagheer, F.A. Non-noble, efficient catalyst of unsupported α-Cr_2O_3 nanoparticles for low temperature CO Oxidation. *Sci. Rep.* **2017**, *7*, 14788. [CrossRef]
44. Yang, X.; Wang, H.; Song, Y.; Liu, K.; Huang, T.; Wang, X.; Zhang, C.; Li, J. Low-Temperature Synthesis of a Porous High-Entropy Transition-Metal Oxide as an Anode for High-Performance Lithium-Ion Batteries. *ACS Appl. Mater. Interfaces* **2022**, *14*, 26873–26881. [CrossRef]
45. Chua, X.J.; Pumera, M. Molybdenum Sulfide Electrocatalysis is Dramatically Influenced by Solvents Used for Its Dispersions. *ACS Omega* **2018**, *3*, 14371–14379. [CrossRef]
46. Resnik, D.; Pecar, B.; Mozek, M.; Lokar, N.; Vrtacnik, D. Formation of thin film Ag/AgCl reference electrode by electrochemical and chemical method. In Proceedings of the 2019 42nd International Convention on Information and Communication Technology, Electronics and Microelectronics (MIPRO), Opatija, Croatia, 20–24 May 2019.
47. Oje, A.M.; Ogwu, A.A. Chromium oxide coatings with the potential for eliminating the risk of chromium ion release in orthopaedic implants. *R. Soc. Open Sci.* **2017**, *4*, 170218. [CrossRef]
48. Meek, R.M.; Afolaranmi, G.A.; Tettey, J.; Grant, M.H. Release of chromium from orthopaedic arthroplasties. *Open Orthop. J.* **2008**, *2*, 10–18. [CrossRef]
49. Eisenbarth, E.; Velten, D.; Müller, M.; Thull, R.; Breme, J. Biocompatibility of β-stabilizing elements of titanium alloys. *Biomaterials* **2004**, *25*, 5705–5713. [CrossRef] [PubMed]

50. Bera, T.K. Bioelectrical Impedance and the Frequency Dependent Current Conduction through Biological Tissues: A Short Review. *IOP Conf. Ser. Mater. Sci. Eng.* **2018**, *331*, 012005. [CrossRef]
51. Fu, Y.; Li, J.; Luo, H.; Du, C.; Li, X. Recent advances on environmental corrosion behavior and mechanism of high-entropy alloys. *J. Mater. Sci. Technol.* **2021**, *80*, 217–233. [CrossRef]

nanomaterials

Review

Principles and Applications of Two-Dimensional Semiconductor Material Devices for Reconfigurable Electronics

Jiong Pan [1,2,†], Yike Zhang [3,†], Jiaju Yin [1,2], Pengwen Guo [1,2], Yi Yang [1,2,*] and Tian-Ling Ren [1,2,*]

1 School of Integrated Circuits, Tsinghua University, Beijing 100084, China
2 Beijing National Research Center for Information Science and Technology (BNRist), Tsinghua University, Beijing 100084, China
3 Weiyang College, Tsinghua University, Beijing 100084, China
* Correspondence: yiyang@tsinghua.edu.cn (Y.Y.); rentl@tsinghua.edu.cn (T.-L.R.)
† These authors contributed equally to this work.

Abstract: With the advances in edge computing and artificial intelligence, the demands of multifunctional electronics with large area efficiency are increased. As the scaling down of the conventional transistor is restricted by physical limits, reconfigurable electronics are developed to promote the functional integration of integrated circuits. Reconfigurable electronics refer to electronics with switchable functionalities, including reconfigurable logic operation functionalities and reconfigurable responses to electrical or optical signals. Reconfigurable electronics integrate data-processing capabilities with reduced size. Two-dimensional (2D) semiconductor materials exhibit excellent modulation capabilities through electrical and optical signals, and structural designs of 2D material devices achieve versatile and switchable functionalities. 2D semiconductors have great potential to develop advanced reconfigurable electronics. Recent years witnessed the rapid development of 2D material devices for reconfigurable electronics. This work focuses on the working principles of 2D material devices used for reconfigurable electronics, discusses applications of 2D-material-based reconfigurable electronics in logic operation and artificial intelligence, and further provides a future outlook for the development of reconfigurable electronics based on 2D material devices.

Keywords: two-dimensional materials; two-dimensional semiconductors; reconfigurable electronics; reconfigurable logic; artificial intelligence

1. Introduction

Over the past decades, the trend of promoting the complementary metal-oxide-semiconductor (CMOS) technology integration level along with the predictions of Moore's Law [1] has been facilitating the boost of computation power and the prosperity of artificial intelligence (AI). However, the rising demands of the data throughput and scaling down with the rapid development of big-data-processing and edge-computing scenarios are increasingly hard to be met, as conventional integrated circuit (IC) technology approaches its physical limit [2,3]. Therefore, extensive works have been conducted for innovative architectures, including devices based on silicon nanowires or nanosheets [4,5], carbon nanotubes [6,7], and two-dimensional (2D) semiconductor materials [8,9].

2D semiconductor materials have drawn increasing attention in recent years. 2D semiconductor materials are free of dangling bonds with ultra-thinned thickness and have less mobility degradation effects than silicon in the same thickness scale. 2D semiconductors with reduced thickness exhibit enhanced control of the channel current when the channel length is down to 10 nm, which promotes the transistor channel length scaling down [10,11]. The ultra-thinned bodies of 2D semiconductors enable easy electrostatic gating control of 2D semiconductor channels and channel polarity tuning between p-, n-, and intrinsic types [12]. 2D materials exhibit prominent physics properties, including confinement effects [13], optical responses [14], etc., which are promising for developing devices with extended functionalities. Large-area growth technologies of 2D semiconductor materials, including chemical vapor deposition [15], have been investigated to promote the applications of 2D semiconductors to large-scale integrated circuits. Recently, transistor structures in advanced nodes, including FinFETs [16] and GAAFETs [17], based on 2D semiconductor materials have been investigated to further enhance gate control, and three-dimensional (3D) integration designs of 2D material devices with higher area efficiency have been developed to validate the feasibility of 2D semiconductors for 3D stacking technology [18–20]. 2D materials are promising for the next-generation computational devices [21,22].

Recently, reconfigurable electronics based on 2D semiconductor materials have been investigated to overcome the physical limit and promote the area scaling down. 2D semiconductor channels with ambipolar characteristics can be easily doped as p-type or n-type channels by the gate voltage, and the channel current control dimension is increased at the device level [23–25]. 2D heterojunctions exhibit physical properties that modulate the transfer or output characteristics of channels and extend the functionalities of devices [26]. Reconfigurable electronics utilizing the properties of 2D semiconductors can change their functionalities to integrate multiple operations into circuit or device components, thereby reducing the number of transistors and increasing area efficiency [27,28]. Reconfigurable logic circuits based on 2D semiconductors have been developed to perform multiple types of logic operations, including basic logic like "AND" and "OR" gates and combinational logic like "multiplexer", by highly simplified circuit forms, which largely reduces the required number of devices needed for reconfigurable logic operations [29–31]. 2D material devices with memory properties can switch their functions between transistor and memory, thereby integrating logic operation and data storage functions into the same device prototype [32]. Apart from digital logic operations, reconfigurable electronics based on 2D material devices have been investigated for analog computing of AI algorithms. Artificial synapses based on reconfigurable 2D material devices have been designed to emulate brain synaptic functions at the device level, which can be used for neuromorphic computing architectures [33]. 2D semiconductor channels with adjustable optical responses can be used to develop in-sensor computing architectures to reduce the data transmission pathway. Those architectures have been used for analog computing of artificial neural networks [34,35].

This work discusses reconfigurable electronics based on 2D semiconductor materials and their recent progress. Section 1 introduces the properties, preparation and growth technologies, and polarity control strategies of 2D semiconductor materials. Section 2 discusses 2D material devices with different mechanisms for reconfigurable functionalities. Sections 3 and 4 analyze recently developed 2D-material-based reconfigurable electronics for logic operations and artificial intelligence, respectively. Finally, the future outlook of the development of reconfigurable electronics based on 2D material devices is given in Section 5.

2. Overview of 2D Semiconductor Materials

2.1. 2D Semiconductor Material Properties

2D semiconductor materials have atomically thin bodies in each layer. Common 2D semiconductor materials include transition metal disulfide (TMDCs) (Figure 1a) [36] and black phosphorus (BP) (Figure 1b) [37]. Compared with bulk materials, 2D semiconductor materials have several prominent physical properties. With the reduction in channel material thickness for bulk materials like silicon, the degradation in mobility is severe due to the state of the surface trap and dangling bonds. On the contrary, the atomic arrangement of ultra-thinned 2D material layers has a pristine interface. The density of trap states is minimized, and scattering at channel–dielectric interfaces is reduced [38]. Therefore, 2D materials maintain good intrinsic mobility in atomical thickness. The reduction in channel thickness leads to the increase of the gate control capability in ultra-scale channel length, so 2D semiconductors with ultra-thinned thickness can be used to achieve short-channel transistors [39]. The thin body also enables easy electrostatic gating control to tune the energy band of the channel and change the polarity of materials [29]. 2D materials have good flexibility and mechanical strength to enable flexible electronics [40,41]. 2D semiconductors like MoS_2 have ultra-high sensitivity in response to optical information [42], and 2D semiconductors can be used for high-performance photoelectric devices, including photodiodes [43] and spectrometers [44]. Two-dimensional semiconductor materials have drawn attention in the development of post-silicon transistor design and manufacturing [45].

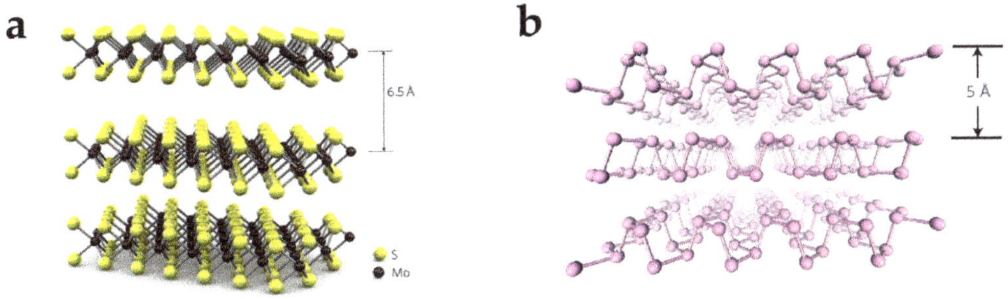

Figure 1. 2D semiconductor structures. (**a**) The structure of MoS_2 as a 2D TMDC material. The single-layer thickness is 6.5 Å. Reproduced with permission [36], copyright 2011, Springer Nature. (**b**) The structure of BP. The single-layer thickness is 5 Å. Reproduced with permission [37], copyright 2014, Springer Nature.

2.2. Preparation and Growth Methods of 2D Semiconductor Materials

Early discoveries and investigations on 2D materials are based on the mechanical exfoliation method [36,46]. Nowadays, mechanical exfoliation is commonly used to fabricate single or small-scale 2D material devices. The steps of mechanical exfoliation are shown in Figure 2a. First, the bulk crystal is pressed onto Scotch tape. Then, the Scotch tape is lifted to take 2D material flakes away from the bulk crystal, and few-layer 2D material flakes are located and transferred to the substrate by polydimethylsiloxane (PDMS). The transferred 2D material layer can be patterned by etching, and other fabrication processing like metal deposition can be conducted in the following steps. Mechanical exfoliation is a straightforward method to obtain thin-layer 2D materials. There are no chemical damage or high-temperature effects, and 2D materials made from mechanical exfoliation have excellent electrical and optical response properties and retain intrinsic physical properties [47].

Since there are problems of large device-to-device variability and hardness to achieve large-area synthesis, other methods have been developed for large-scale integration of 2D semiconductors. CVD is regarded as one of the most promising methods for the synthesis of large-area and single-layer two-dimensional materials [48]. CVD is an atomic-level surface modification process where a thin solid film is deposited on an underlying substrate through a chemical reaction from the vapor or gas phase. It usually has two steps, mass transport and a surface kinetics reaction [49]. CVD includes various branch methods, such as metal–organic CVD (MOCVD), plasma-enhanced CVD (PECVD), low-pressure CVD (LPCVD), etc. The schematic diagram of MOCVD is shown in Figure 2b. The high temperature used in CVD enables the high migration ability of atoms and molecules and promotes the growth of crystalline domains, which results in large domain sizes and good uniformity [50]. Main stream CVD growth methods have to use high temperatures to achieve good crystalline, while back-end-of-the-line technologies have maximum temperatures of about 500 °C [51]. A widely used solution is to transfer 2D material layer fabricated by CVD on a sapphire substrate to the chip. The transfer method of CVD 2D materials can be used for back-end-of-the-line technologies for wafer-scale production. Low-thermal-budget CVD synthesis approaches have also been developed in recent years [52]. Apart from mechanical exfoliation and CVD, other methods, including liquid phase exfoliation [53] and epitaxial growth [54], have been also used for the preparation and growth of 2D semiconductors.

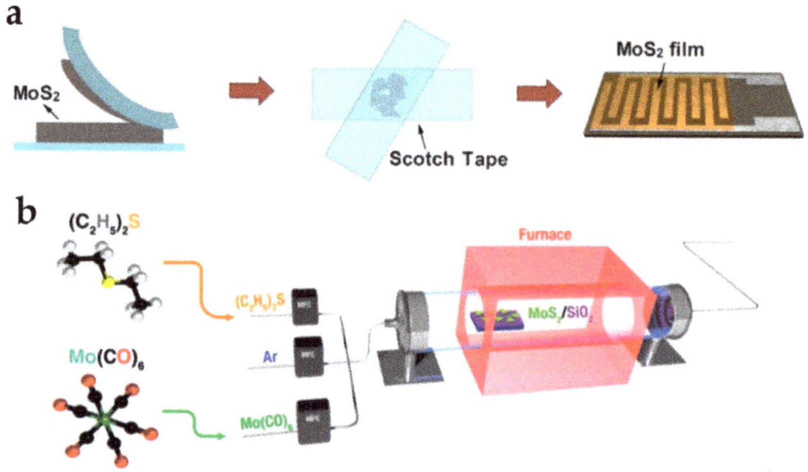

Figure 2. 2D semiconductor fabrication methods. (**a**) Steps of mechanical exfoliation. Reproduced with permission [55], copyright 2019, MDPI AG. (**b**) Schematic diagram of MOCVD. Reproduced with permission [56], copyright 2021, Wiley-VCH.

2.3. Polarity of 2D Semiconductor Materials

Due to versatile physical properties of 2D semiconductors, different 2D semiconductor materials exhibit n-type, p-type, or ambipolar transport behaviors, thereby promoting the energy band engineering, and can achieve various operation functionalities. As shown in Figure 3a,b, the n-type MoS_2 channel can be used for "OR" logic operation, the p-type BP channel can be used for "NOR" operation, and the ambipolar WSe_2 channel can achieve "XNOR" logic operation [57]. Proper designs of device structures based on 2D semiconductors with different polarity properties can develop various reconfigurable electronics.

Polarities of 2D semiconductors, including WSe_2 [58], $MoTe_2$ [59], and BP [60], can be easily tuned since 2D semiconductors have ultra-thinned bodies. 2D semiconductor layers can exhibit ambipolar, unbalanced ambipolar, p-type, or n-type transport behaviors [61–63]. Various strategies can be applied to tune the mobilities and carrier densities of electrons and holes and change the semiconductors. For example, the atomic layer deposition (ALD) of high-k dielectric, including Al_2O_3, can enhance n-type behavior (Figure 3c,d).

To promote the applications of 2D materials in integrated circuits and very-large-scale integration, the performance of 2D semiconductor channels with both p- and n-type polarities should be enhanced to build complementary circuits. The development progress of high-performance n-type 2D channels like CVD-grown MoS_2 is relatively advanced [64,65], but the development of pure p-type 2D channels is limited. There exits Fermi-level pinning near the conduction band caused by the interfacial defects induced by evaporation of the contact metal. Strategies to obtain high-performance p-type 2D channels have been developed in recent years by the stable evaporation of high-work-function metals [66]. Complementary circuits with n-type and p-type 2D material channels for 2D integrated circuits have been developed recently (Figure 3e,f) [67].

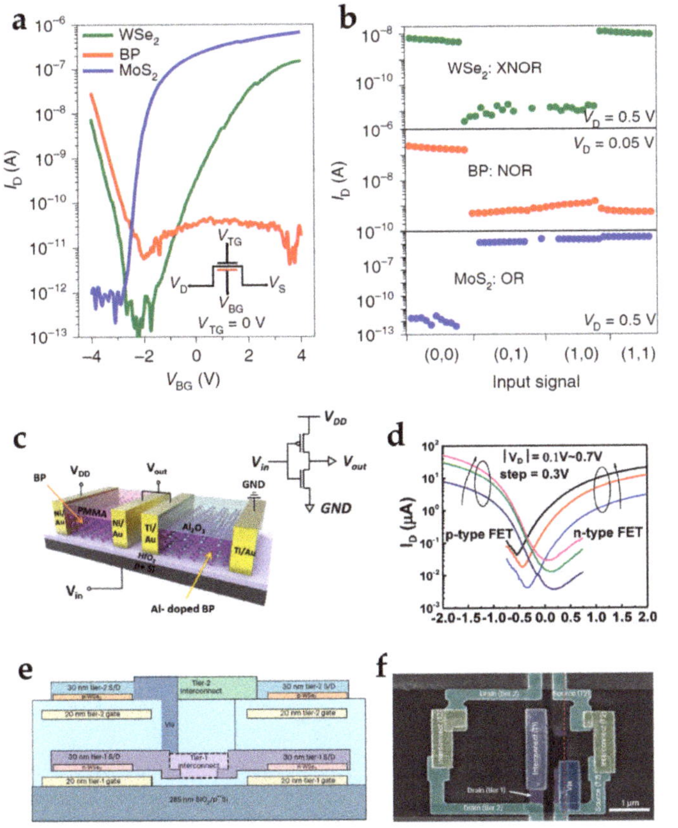

Figure 3. The polarity control of 2D semiconductor materials. (**a**) Different polarity properties of 2D semiconductor materials and (**b**) their corresponding operation applications. Reproduced with permission [57], copyright 2021, Springer Nature. (**c**,**d**) BP transistors with tunable polarities by the Al-donor doping technique of Al_2O_3 ALD processes. (**c**) The inverter with p-type BP and n-type Al-doped BP transistors. Reproduced with permission [62], copyright 2018, Wiley-VCH. (**d**) Transport curve of p-type and n-type BP transistors. Reproduced with permission [62], copyright 2018, Wiley-VCH. (**e**) The schematic and (**f**) scanning electron microscopy image of the complementary circuits composed of p-WSe2 and n-WSe2 by evaporations of contact metal with different work functions. Reproduced with permission [67], copyright 2024, Springer Nature.

3. Working Principle of Reconfigurable Devices

Various strategies have been investigated for the implementation of reconfigurable devices to achieve different functionalities, including homojunction, heterojunction, and defects. Homojunction is formed by a 2D semiconductor material layer, which is more easily fabricated, and has fewer defects at the junction interface. The positions of junctions and the positions of gate boundaries are self-aligned. Heterojunctions are formed by the stacking of 2D materials. With proper structural designs, the modulation functions of heterojunctions can be tuned by a single gate. Heterojunctions deliver various physical properties due to the interactions of multiple 2D semiconductors. Defects in 2D material layers or 2D material device structures induce a charge trapping/de-trapping effect to implement programming or memory functionalities of reconfigurable electronics.

3.1. Homojunction

A tunable homojunction can change the relations between the gate voltage and the drain current and therefore change the functionalities of devices. p–n homojunctions are widely used in 2D-material-based reconfigurable devices. A p–n homojunction with switchable polarities for ambipolar 2D materials can be used to build structures or circuits with various reconfigurable functions. A switchable p–n homojunction can be applied to device structures for circuits with reconfigurable logic computing (Figure 4a,b) [68,69]. Since photocurrents can be generated at p–n junctions, a switchable p–n homojunction can also change the direction of a photocurrent induced by light information and can be used for reconfigurable optoelectronic devices [70,71]. The doping of 2D materials can be conducted by ion implantation [72,73], but the polarity is unchanged after doping. Due to the ultra-thin body of 2D materials, electrostatic gating is an effective strategy to tune the energy band in a sufficiently large range and can dynamically change the polarity of the gate-controlled channel [12]. Other strategies for a switchable p–n homojunction have also been investigated. For example, Peng et al. proposed a single-gate reconfigurable device based on graded doping by the absorption and desorption of gas molecules (Figure 4c) [74]. The graded doping strategy can form a p–n homojunction for ambipolar $MoTe_2$ with a single gate, which simplifies the device complexity (Figure 4d). Apart from a p–n homojunction, junctions with the same polarities, including n^+–n^- or p^+–p^- homojunctions, can also be used to achieve reconfigurable functions [75]. An n^+–n^- or p^+–p^- homojunction achieves various functionalities with less demand for the energy band tuning range. The integration of multiple forms of junctions can further enhance the functionality of reconfigurable structures. Pan et al. developed a multibarrier collaborative modulation architecture (Figure 4e,f) through interactions of p–n, n^+–n^-, and p^+–p^- homojunctions (Figure 4g) and integrated combinational and reconfigurable logic computing functions at the device level [31].

3.2. Heterojunction

Heterojunctions have versatile electric modulation properties that can be applied to reconfigurable electronics. The unequal band gaps between two semiconductor materials induce more control dimensions for the transfer or output characteristics of devices. A commonly applied property is negative differential resistance (NDR). The energy bands of channels are variable by gate voltages, and the potential differences of the valence bands and conduction bands between two materials at the heterojunction are tunable [76]. When the tunable range is sufficiently large, the polarities of the potential differences can be altered, which can change the type of the major carrier in the channels or switch the relation between gate voltage and the carrier density [77,78]. Therefore, the curves of I–V characteristics exhibit different polarities of derivatives in different voltage intervals. 2D heterojunctions can be used for electronics with multi-valued logic operations. Huang et al. proposed a multifunctional device based on a BP/MoS_2 heterojunction (Figure 5a) [79]. The NDR transfer characteristics enable the device to be used for a reconfigurable inverter that can process binary and ternary logic controlled by the supply voltage (Figure 5b). Types of logic operation functions can also be reconfigured by 2D heterojunctions. Seo et al. presented a BP/ReS_2 heterojunction reconfigurable device [80]. NDR property is induced by the type-III BP/ReS_2 heterojunction, and electron trapping/de-trapping is induced to tune the layer resistance. Reconfigurable circuits for ternary inverter and latch are built by the heterojunction. Another widely used property of a 2D heterojunction is negative differential transconductance or anti-ambipolar characteristics. The current is relative large for gate voltage that at the middle of the interval, and when increasing or decreasing the gate voltage, the current is reduced [81]. Shingaya et al. proposed a dual-gate anti-ambipolar reconfigurable device based on a ReS_2/WSe_2 heterojunction (Figure 5c,d) [82]. The anti-ambipolar device can achieve two-input logic operations of AND, OR, XOR, NAND, NOR, and XNOR (Figure 5e).

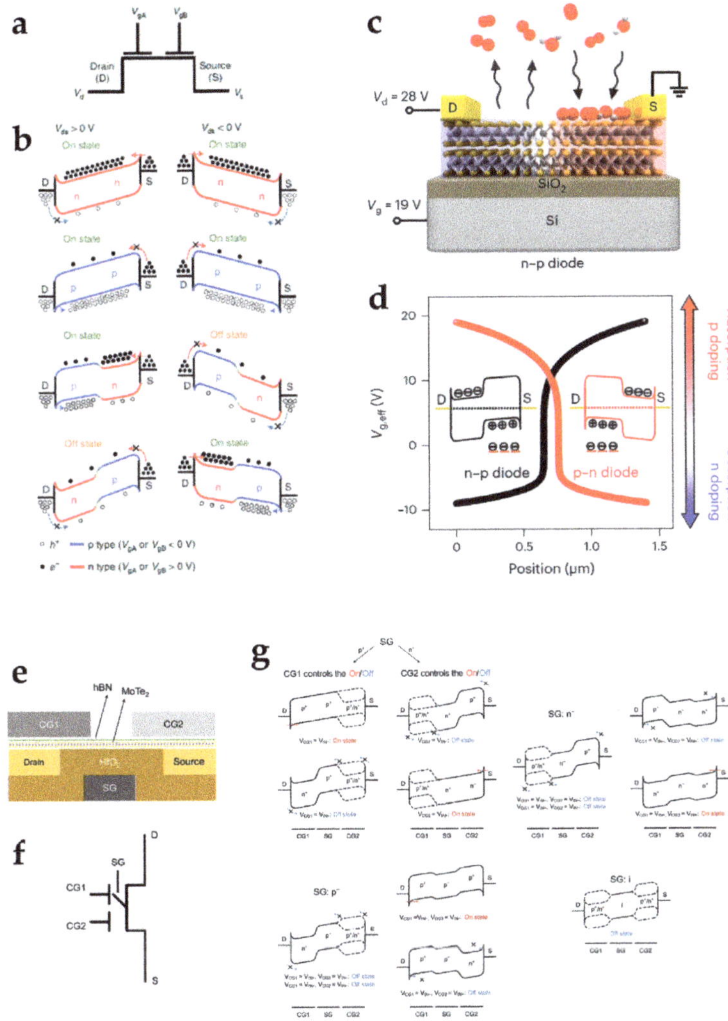

Figure 4. 2D-material-based reconfigurable electronics based on homojunction structures. (**a**) The schematic symbol and (**b**) band diagrams of the electrically tunable two-dimensional homojunction for reconfigurable logic computing. Reproduced with permission [29], copyright 2020, Springer Nature. (**c**) The device schematic and (**d**) polarity switchable characteristics of the single-gate 2D material device implementing reconfigurable functions by graded doping. Reproduced with permission [74], copyright 2023, Springer Nature. (**e**) The device structure, (**f**) schematic symbol, and (**g**) band diagrams of the multibarrier collaborative modulation device. Reproduced with permission [31], copyright 2024, American Chemical Society.

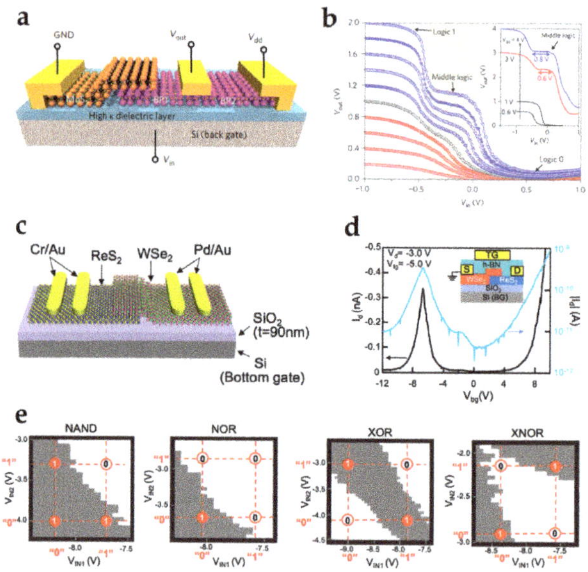

Figure 5. 2D-material-based reconfigurable electronics based on heterojunctions. (**a**) A multifunctional BP/MoS_2 heterojunction-based device for (**b**) tunable multi-value logic operations. Reproduced with permission [79], copyright 2017, Springer Nature. (**c**) A ReS_2/WSe_2 heterojunction-based reconfigurable device exhibiting (**d**) anti-ambipolar characteristics. Reproduced with permission [82], copyright 2022, Wiley-VCH. (**e**) Logic operations of the anti-ambipolar reconfigurable device to perform "NAND", "NOR", "XOR", and "XNOR" logic. Reproduced with permission [82], copyright 2022, Wiley-VCH.

3.3. *Charge Trapping/De-Trapping of Defects*

Defects in 2D material devices have been utilized in reconfigurable electronics. In 2D material layers, defects refer to structural abnormalities in a material due to the absence or irregular arrangement of atoms such as atomic vacancies or heteroatoms [83]. The defects of 2D TMDs are often expressed as intrinsic sulfur vacancies [84]. Defects of 2D material layers significantly affect the materials' physical properties, including optical characteristics [85], electrical [86] and thermal [87] conductivity, etc. A widely applied design of utilizing the defects of 2D material layers is the control of charge trapping/de-trapping to or from the 2D material layer defects (Figure 6a) [80]. The 2D material layer conductivity is thereby adjusted (Figure 6b), and the I–V characteristics curves are tunable to achieve reconfigurable functionalities. Apart from defects in 2D material layers, charge trapping/de-trapping processes can be generated by interfaces of 2D material devices. Tsai et al. proposed a reconfigurable device based on a hBN/$ReSe_2$/hBN heterostructure (Figure 6c) [32]. The non-volatile programmable functions of the device were based on the defect states in hBN and photoinduced trapping at the hBN/SiO_2 interface. Under the control of the program gate voltage, holes or electrons tunneled through the hBN layer when the light signal was on and were trapped at the hBN/SiO_2 interface (Figure 6d). The trapped charge exhibited non-volatile behavior when the light signal was off and electrostatically tuned the polarization states of the $ReSe_2$ channel.

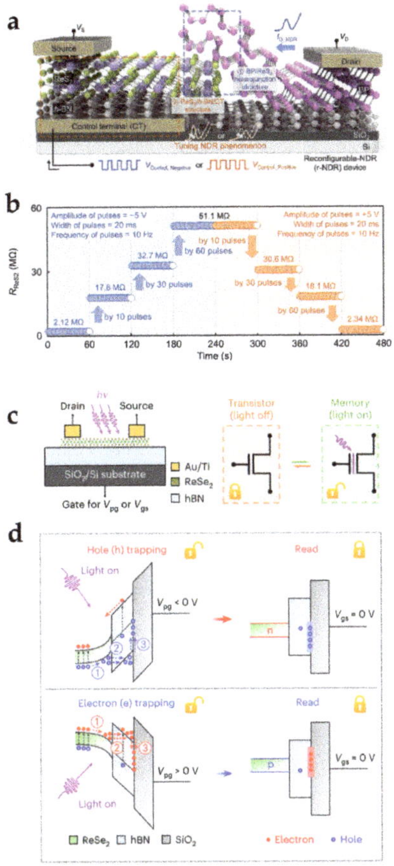

Figure 6. Defect functionalities in reconfigurable 2D material devices. (**a**) A reconfigurable device exhibiting tunable channel resistance by the charge trapping/de-trapping processes to or from vacancies in the ReS_2 layer. Reproduced with permission [80], copyright 2022, Wiley-VCH. (**b**) The resistance change of the ReS_2 layer by charge trapping/de-trapping under positive or negative electrical pulses. Reproduced with permission [80], copyright 2022, Wiley-VCH. (**c**) A reconfigurable transistor and memory with photoinduced trapping property at the dielectric interface. Reproduced with permission [32], copyright 2023, Springer Nature. (**d**) A schematic of the photoinduced trapping at the hBN/SiO_2 interface. Reproduced with permission [32], copyright 2023, Springer Nature.

4. Two-Dimensional Semiconductor Reconfigurable Devices for Logic Operations

Recent research on logic devices based on two-dimensional semiconductor materials is of various functions and application scenarios. Reconfigurable logic devices have advantages, including area saving, low power consumption, etc. This section discusses the applications of 2D-material-based reconfigurable electronics for logic circuits.

4.1. Reconfigurable Logic Operation Circuits

Designs of reconfigurable electronics based on 2D material devices are investigated to promote the functionalities of reconfigurable logic circuits. Pan et al. proposed an electrically tunable homojunction device based on WSe_2. The direction of the p–n junction and the drain current can be controlled by a combination of gate and drain voltage input polarity. When the polarities of the gate voltages were the same, the device channel appeared as n–n- or p–p-type doping. When the polarities were opposite, n–p or p–n junctions were formed, and multiple different current states were achieved. The device was used to build modules that have reconfigurable logic operation functions of two-input logic operations and multiplexer, d-latch, and adder/subtractor [29]. Wu et al. proposed a reconfigurable transistor based on a BP channel to build complementary logic gates (Figure 7a) [12]. This device exhibited programmable p-type and n-type transistor modes by adjusting the polarity gate voltage. Logic circuits were achieved to perform reconfigurable functions of "NAND/NOR" or "XOR/XNOR" operation functions (Figure 7b). Apart from increasing circuit functionalities, designs of 2D-material-based reconfigurable devices to increase area efficiency are also developed. Pan et al. reported a multibarrier collaborative modulation device for device-level high-density reconfigurable logic computing. Interactions of multiple forms of potential barriers were introduced to enhance the logic operation functionality at the device level. A percentage of 58.8% and 71.4% area is saved for combinational and reconfigurable logic operations [31]. To meet the demand for greater power saving and the reduction of system complexity in binary logic circuits, multi-valued logic circuits are attracting attention [88,89]. Yi et al. proposed a double-gate transistor architecture based on a MoS_2 homojunction (Figure 7c) [90]. In the double-gate structure, the combination of top gate and bottom gate voltages controlled the charge-doping states of the channel. By adjusting the voltages, the charge-doping state had depletion, neutral, and accumulation modes. The device can be used as a circuit that can change the function between binary and ternary logic inverters by a gate voltage (Figure 7d).

4.2. Reconfigurable Logic and Memory Circuits

Strategies to integrate multiple-data-processing functionalities have been realized. Devices integrating logic operation and non-volatile memory were developed. Sun et al. developed a reconfigurable logic-in-memory architecture based on an ambipolar WSe_2 homojunction and graphene partial floating gate (Figure 8a) [91]. The partial floating gate was used to store charge under specific voltage conditions. Homojunctions were formed between the channels controlled by the partial floating gate and the top gate, and the on/off states of the channel current were determined by the equality between the stored charge polarity and the top gate voltage. Reconfigurable circuits were developed to perform logic-in-memory operations of two-input logic (Figure 8b). Zeng et al. reported a side-gate reconfigurable device based on a MoS_2 channel and hBN dielectric (Figure 8c,d) [92]. The conductivity of the channel was selectively regulated by side-gate voltages. Charge storage and erase at the top Au floating gate were implemented to achieve non-volatile memory. The device performed multiple functionalities, including diode, reconfigurable logic transistor, and floating-gate memory functions. Designs for the integration of multi-valued logic operations and memory functions were also proposed. Wang et al. proposed a semi-floating-gate-controlled 2D InSe homojunction (Figure 8e) [93]. A multilayer graphene floating gate performed charge storage. By applying a voltage pulse to the control gate of the silicon substrate, charge carriers tunneling through the hBN layer changed the carrier polarities of the charge stored in the multilayer graphene floating gate, forming different types of junctions in the InSe channel (Figure 8f). The structure implemented dynamic conversion between logic rectifiers, memories, and multi-valued logic inverters. Non-volatile memory functions have also been used for device programming to extend the logic operation functionalities of reconfigurable devices. Seo et al. proposed a ternary logic reconfigurable device latch based on a BP/ReS_2 heterojunction [80]. Ternary logic operation is implemented by the NDR characteristics of the heterojunction, and electron trapping/de-trapping effects at ReS_2 layer interfaces are utilized for logic function reconfiguration between ternary inverter and ternary latch operations.

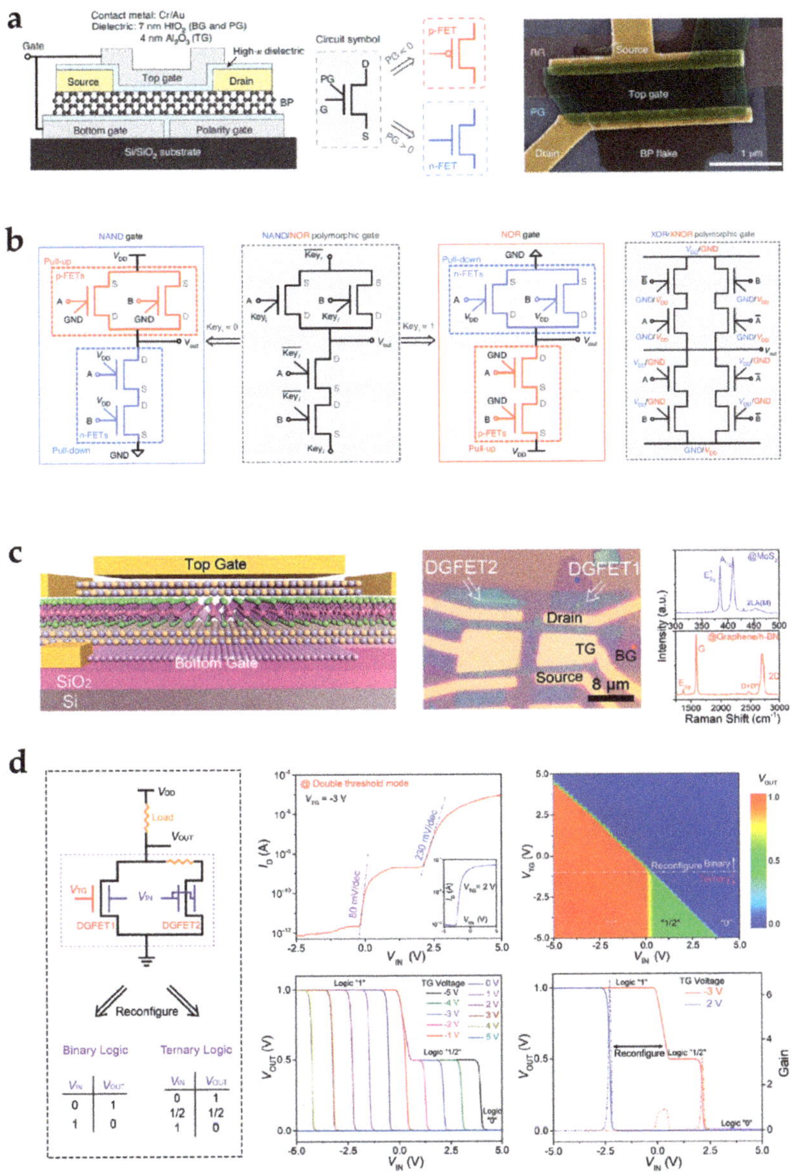

Figure 7. 2D-material-based reconfigurable electronics for logic operation circuits. (**a**) Device characterizations and (**b**) NAND/NOR and XOR/XNOR polymorphic gates of the double-gate BP Schottky-barrier field effect transistor and circuits. Reproduced with permission [12], copyright 2020, Springer Nature. (**c**) Device characterizations and (**d**) logic functions of the double-gate transistor based on a MoS_2 homojunction performing reconfigurable binary and ternary inverter circuit. Reproduced with permission [90], copyright 2021, Wiley-VCH.

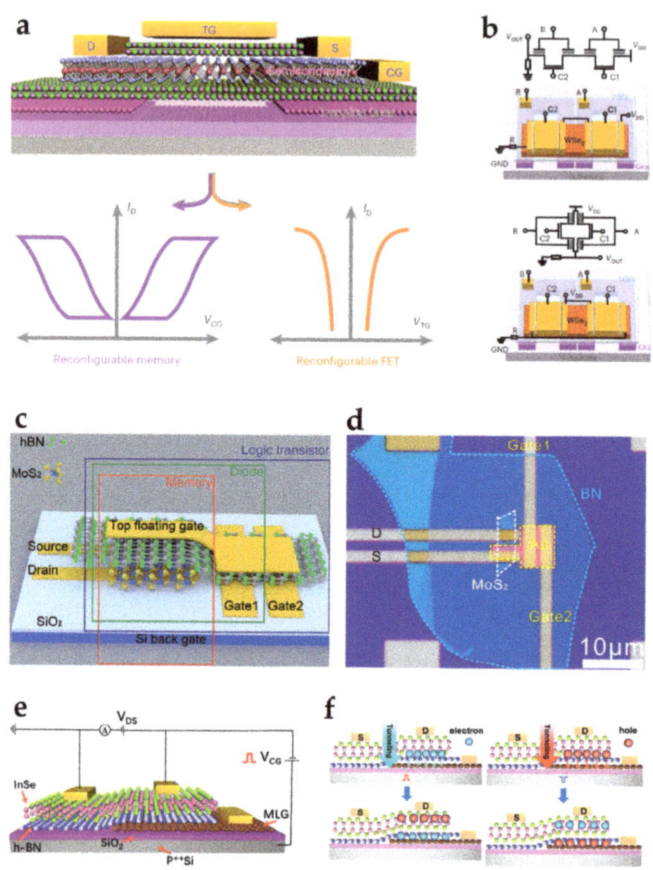

Figure 8. 2D-material-based reconfigurable electronics for reconfigurable logic and memory circuits. (**a**) A partial-floating-gate architecture and (**b**) reconfigurable logic-in-memory circuits. Reproduced with permission [91], copyright 2022, Springer Nature. (**c**) The device schematic and (**d**) optical image of the side-gate BN-MoS_2 transistor for reconfigurable operations of diode, reconfigurable logic, and memory. Reproduced with permission [92], copyright 2023, Wiley-VCH. (**e**) A semi-floating-gate transistor based on InSe homojunction with reconfigurable functions of rectifier, memory device, and ternary inverter. Reproduced with permission [93], copyright 2023, Wiley-VCH. (**f**) A schematic of electron and hole tunneling by voltage pulses. Reproduced with permission [93], copyright 2023, Wiley-VCH.

4.3. *Reconfigurable Optoelectronic Logic Circuits*

Emerging optoelectronic technologies have made it possible for 2D semiconductor reconfigurable devices to be applied in the field of optoelectronic logic and memory devices. Ma et al. reported an optoelectronic reconfigurable logic device based on vertical field-effect transistors with a graphene/MoS_2/WSe_2/graphene heterojunction (Figure 9a) [94]. The device exhibited a reconfigurable modulation of photoresponse by gate and drain voltages. Optoelectronic reconfigurable logic gates of "XNOR", "NOR", "NAND", "AND", "OR", and "Inhibit" were achieved (Figure 9b). Apart from defining optical responses as the outputs of logic gates, optical signals can also be the input of logic gates, where the optical signal and electrical signal are two of the logic gate inputs, and logic operations are implemented by combining the optical and electrical control to the device channel current. Bach et al. proposed a reconfigurable floating-gate optoelectronic memory that can be optically and electrically programmed by laser pulses and gate voltage pulses, respectively (Figure 9c) [95]. Based on the combined electrically and optically controllable charge storage properties, reconfigurable optical logic circuits, including "AND" and "OR" gates, were achieved, where laser and gate voltage pulses were two of the logic inputs, and the channel current was the output signal (Figure 9d). Optoelectronics logic operations are compatible with other functionalities, including optoelectronic artificial synaptic computing. A design to combine reconfigurable optoelectronics logic operations of "AND" and "OR" gates with optoelectronic artificial synapses was investigated to achieve a higher function integration of reconfigurable electronics [96].

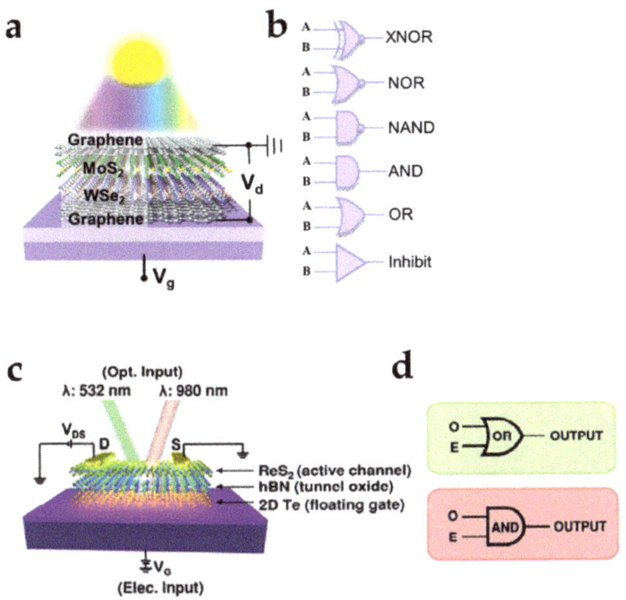

Figure 9. 2D-material-based reconfigurable electronics for optoelectronic logic circuits. (**a**) A graphene/MoS_2/WSe_2/graphene heterojunction device for (**b**) optoelectronic reconfigurable logic gates. Reproduced with permission [94], copyright 2023, American Chemical Society. (**c**) A ReS2/hBN/2D Te floating-gate optoelectronic memory device for (**d**) optoelectronic logic operations of "OR" and "AND" logic. Reproduced with permission [95], copyright 2024, American Chemical Society.

5. 2D Semiconductor Reconfigurable Electronics for Artificial Intelligence

Reconfigurable electronics based on 2D materials have integrated functionalities and are compatible with the analog operations of AI algorithms. Therefore, extensive work has been conducted to develop reconfigurable electronics for AI operations to perform artificial synaptic computing or in-memory and in-sensor computing neural networks. For artificial synapse implementations, 2D material devices respond to the stimuli and output spike and pulse signals to emulate the electric signals of synapses. The output signals can be varied by the reconfiguration of the devices to emulate the variation of inhibitory and excitatory states of synapses. Artificial synapses can be integrated as an operation system to implement neuromorphic computing networks, including spiking neural networks. For in-memory and in-sensor computing neural network implementations, crossbar arrays composed of 2D material devices perform matrix-vector multiplications to implement analog computing. The states of the devices can be independently reconfigured in order to program the weight values of each cell in the network. This section analyzes the recent advances in 2D-material-based reconfigurable electronics for artificial synapses and in-memory/in-sensor computing neural networks.

5.1. 2D-Material-Based Reconfigurable Electronics for Artificial Synapses

Artificial synapses aim to simulate the structure and function of biological neurons to process information in a parallel, pulse-driven manner, and 2D materials are promising for the hardware implementation of basic units of synapses [97,98]. Sahu et al. proposed a MoS_2 optoelectronic artificial synapse based on a photoelectric cooperative stimulation mechanism (Figure 10a) [96]. The light pulse with different wavelengths was irradiated on the monolayer MoS_2 material, which activated the photogenerated electron–hole pairs as charge carriers. Gate voltage was introduced to modulate the transport characteristics of charge carriers. Through photoelectric cooperative stimulation, this device imitated multiple biological neuromorphic behaviors, including long-term potentiation (LTP), and was used as neuromorphic computing architecture units. Hu et al. proposed a reconfigurable neuromorphic unit based on MoS_2/Gr/hBN heterostructure, which emulated the function of key neural elements of the synapse, neuron, and dendrite (Figure 10b) [99]. The function of synapse and dendrite were stimulated by optical stimulation on a MoS_2 photosensitive layer. The graphene electrode and the back gate electrode were used to receive excitatory and inhibitory input signals at the same time. The function of the neuron was stimulated by the Ag filament silver through the hBN layer to imitate the integrate-and-fire behavior of biological neurons, which showed the spike-based information coding ability. Yao et al. proposed a reconfigurable artificial synapse with a WSe_2 channel and $MoTe_2$ floating gate (Figure 10c) [100]. By adjusting the control voltage, the charge states in the floating gate were changed. The device exhibited properties of excitatory and inhibitory postsynaptic current, paired-pulse facilitation, and long-term potentiation and depression to emulate the functionalities of biological synapses.

5.2. 2D-Material-Based Reconfigurable Electronics for In-Memory and In-Sensor Computing Neural Networks

The rapid development of artificial intelligence and edge computing brings growing demands for computational electronics. Conventional von Neumann architecture applied to most computational electronics is confronted with large time latency and power consumption due to the redundant pathway for information sensing, conversion, transmission, storage, and computing [101–103]. To address this problem, 2D material devices with reconfigurable functionalities have been investigated to enhance the performance of neural network processing by combining the functionalities of sensing, memory, and computing. Mennel et al. proposed an analog vision sensor based on a WSe_2 homojunction (Figure 11a) [34]. The p–n homojunction direction was tunable with gate voltages, and the photocurrent direction can be programmed through gate voltage configuration. An analog computing architecture for image processing was established, and the directions of photocurrents were the weights of the neural network. The vision sensor implemented ultra-fast image processing with 40 ns. Wu et al. further extended the device-level functionalities of reconfigurable electronics and integrated sensing and memory functions into the device level through the $MoTe_2$ layer with an organic ferroelectric poly(vinylidene fluoride) and trifluoroethylene (P(VDF-TrFE)) layer (Figure 11b) [35]. An in-memory sensing and computing architecture was developed. The applications of 2D-material-based reconfigurable electronics for in-memory and in-sensor computing have also been extended from static object recognition to motion detection. Zhang et al. proposed a 2D retinomorphic hardware based on a BP channel and WSe_2 floating gate (Figure 11c) [104]. Based on the photoconductivity properties of WSe_2, the device generated adjustable positive and negative responses to light stimuli and achieved moving target detection through inter-frame difference calculation.

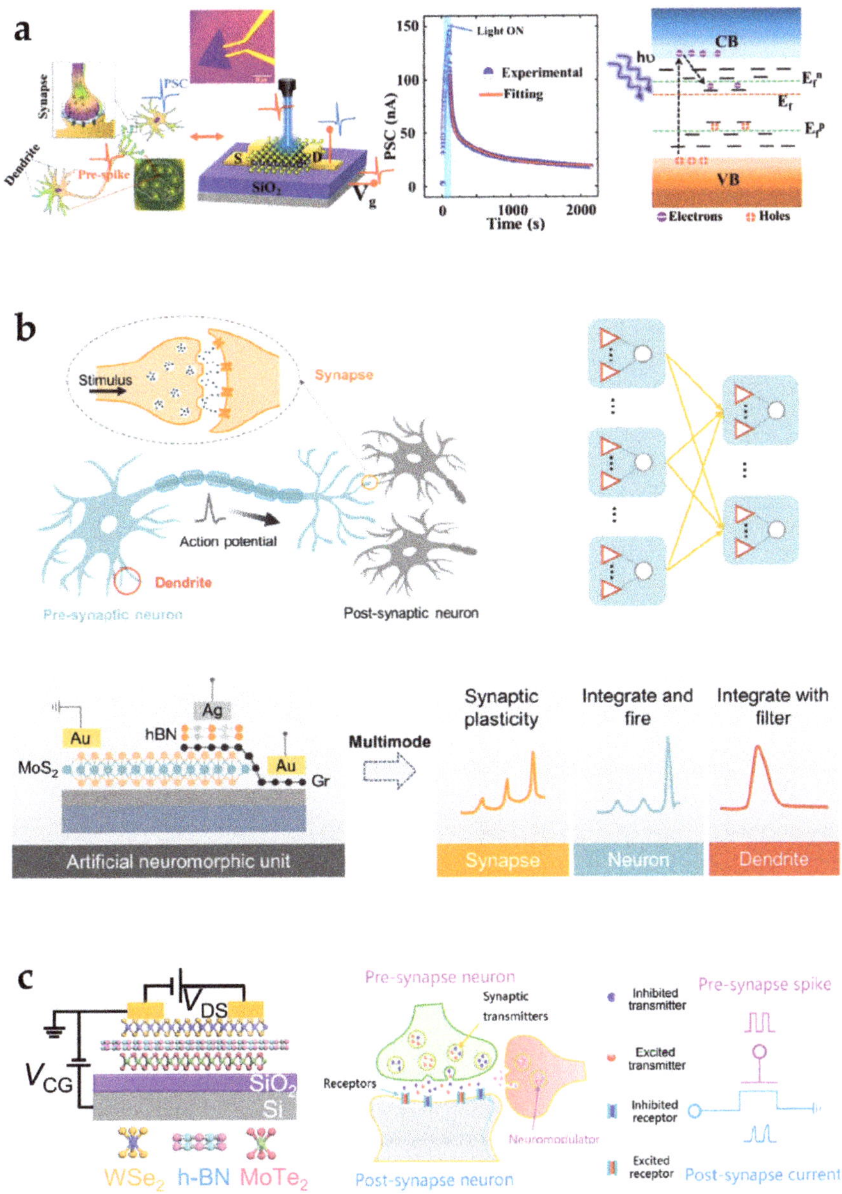

Figure 10. Artificial synapses based on 2D-material-based reconfigurable electronics. (**a**) 2D MoS_2 artificial synapses for in-memory neuromorphic computing. Reproduced with permission [96], copyright 2022, Wiley-VCH. (**b**) Reconfigurable neuromorphic computing devices with multiple neural-information-processing functions. Reproduced with permission [99], copyright 2024, American Chemical Society. (**c**) A WSe_2/hBN/$MoTe_2$ artificial synapse emulating reconfigurable excitatory and inhibitory synaptic plasticity. Reproduced with permission [100], copyright 2023, American Chemical Society.

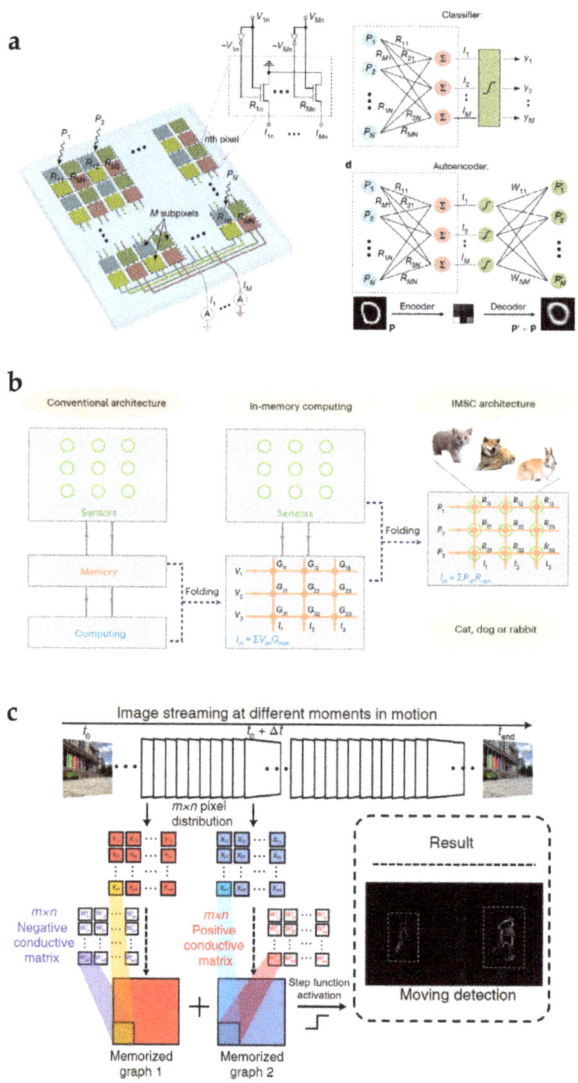

Figure 11. In-memory and in-sensor computing based on reconfigurable 2D material device. (**a**) Ultrafast vision sensor based on WSe_2 homojunction. Reproduced with permission [34], copyright 2020, Springer Nature. (**b**) In-memory sensing and computing neural network based on $MoTe_2$ homojunction and (P(VDF-TrFE)) ferroelectric layer. Reproduced with permission [35], copyright 2023, Springer Nature. (**c**) All-in-one 2D retinomorphic device for motion recognition based on BP channel and WSe_2 floating gate. Reproduced with permission [104], copyright 2022, Springer Nature.

6. Conclusions and Outlooks

Reconfigurable electronics can be applied to promote the function integration level and extend the data-processing functionalities for hardware performing logic operations and AI algorithms. 2D semiconductor materials can be used to develop high-performance reconfigurable electronics. This work reviews the basic properties of 2D materials, principles of 2D material devices for reconfigurable functionalities, and applications of 2D reconfigurable electronics. Presently, multiple structural designs of 2D-material-based reconfigurable electronics have been reported. The mainly utilized fundamental mechanisms of present reconfigurable electronics include homojunctions, heterojunctions, and the charge trapping/de-trapping of defects. In order to extend the functionalities of reconfigurable electronics, future research should exploit further physical properties of 2D semiconductors that have the potential to be used for function reconfiguration. One of the promising properties is the quantum effects of 2D semiconductors that may help to develop low-power or high-speed reconfigurable electronics. Reconfigurable 2D material devices have been investigated to be used for digital operations, including high-density reconfigurable logic circuits and hardware security modules and analog operations, including neuromorphic computing or in-sensor memory and computing. Research works have predicted the advanced performance of 2D-material-based reconfigurable electronics, including large area efficiency and low power consumption. Most presently proposed device or circuit prototypes were fabricated on relatively large sizes. In the future, fabrications of 2D-material-based reconfigurable electronics in advanced technology nodes should be developed to validate the performance advantages and the potential in practical applications. To achieve this objective, the stabilities of reconfigurable operations need to be further increased to meet the requirement of practical applications, and the variability of device units and different cycles should be reduced. Most present designs of reconfigurable 2D material devices are based on mechanical exfoliation. Large-area fabrication strategies of 2D-material-based reconfigurable electronics should be proposed to realize large-scale reconfigurable circuits.

Author Contributions: Writing—original draft preparation, J.P. and Y.Z.; writing—review and editing, J.P. and Y.Z.; visualization, J.Y. and P.G.; supervision, T.-L.R.; funding acquisition, T.-L.R. and Y.Y. All authors have read and agreed to the published version of the manuscript.

Funding: This study was supported by the National Key R&D Program (2022YFB3204100, 2021YFC3002200), the National Natural Science Foundation (U20A20168) of China, the Research Fund from Tsinghua University Initiative Scientific Research Program, and a grant from the Guoqiang Institute, Tsinghua University.

Data Availability Statement: Not applicable.

Conflicts of Interest: The authors declare no conflicts of interest.

References

1. Moore, G.E. Cramming more components onto integrated circuits. *Electronics* **1965**, *38*, 114. [CrossRef]
2. Cao, W.; Bu, H.; Vinet, M.; Cao, M.; Takagi, S.; Hwang, S.; Ghani, T.; Banerjee, K. The future transistors. *Nature* **2023**, *620*, 501–515. [CrossRef] [PubMed]
3. Liu, A.; Zhang, X.; Liu, Z.; Li, Y.; Peng, X.; Li, X.; Qin, Y.; Hu, C.; Qiu, Y.; Jiang, H.; et al. The Roadmap of 2D Materials and Devices Toward Chips. *Nano-Micro Lett.* **2024**, *16*, 119. [CrossRef]
4. Mertens, H.; Ritzenthaler, R.; Hikavyy, A.; Kim, M.S.; Tao, Z.; Wostyn, K.; Chew, S.A.; De Keersgieter, A.; Mannaert, G.; Rosseel, E.; et al. Gate-all-around MOSFETs based on vertically stacked horizontal Si nanowires in a replacement metal gate process on bulk Si substrates. In Proceedings of the 2016 IEEE Symposium on VLSI Technology, Honolulu, HI, USA, 14–16 June 2016; pp. 1–2.
5. Loubet, N.; Hook, T.; Montanini, P.; Yeung, C.W.; Kanakasabapathy, S.; Guillom, M.; Yamashita, T.; Zhang, J.; Miao, X.; Wang, J.; et al. Stacked nanosheet gate-all-around transistor to enable scaling beyond FinFET. In Proceedings of the 2017 Symposium on VLSI Technology, Kyoto, Japan, 5–8 June 2017; pp. T230–T231.
6. Franklin, A.D.; Hersam, M.C.; Wong, H.P. Carbon nanotube transistors: Making electronics from molecules. *Science* **2022**, *378*, 726–732. [CrossRef]
7. Zhang, Z.; Passlack, M.; Pitner, G.; Natani, S.; Su, S.-K.; Chao, T.-A.; Liew, S.L.; Hou, V.D.H.; Hsu, C.-F.; Shipley, W.E.; et al. Complementary carbon nanotube metal–oxide–semiconductor field-effect transistors with localized solid-state extension doping. *Nat. Electron.* **2023**, *6*, 999–1008. [CrossRef]
8. Wachter, S.; Polyushkin, D.K.; Bethge, O.; Mueller, T. A microprocessor based on a two-dimensional semiconductor. *Nat. Commun.* **2017**, *8*, 14948. [CrossRef] [PubMed]

9. Jayachandran, D.; Pendurthi, R.; Sadaf, M.U.K.; Sakib, N.U.; Pannone, A.; Chen, C.; Han, Y.; Trainor, N.; Kumari, S.; Mc Knight, T.V.; et al. Three-dimensional integration of two-dimensional field-effect transistors. *Nature* **2024**, *625*, 276–281. [CrossRef]
10. Shen, Y.; Dong, Z.; Sun, Y.; Guo, H.; Wu, F.; Li, X.; Tang, J.; Liu, J.; Wu, X.; Tian, H.; et al. The Trend of 2D Transistors toward Integrated Circuits: Scaling Down and New Mechanisms. *Adv. Mater.* **2022**, *34*, e2201916. [CrossRef]
11. English, C.D.; Smithe, K.K.H.; Xu, R.L.; Pop, E. Approaching ballistic transport in monolayer MoS_2 transistors with self-aligned 10 nm top gates. In Proceedings of the 2016 IEEE International Electron Devices Meeting (IEDM), San Francisco, CA, USA, 3–7 December 2016; pp. 5.6.1–5.6.4.
12. Wu, P.; Reis, D.; Hu, X.S.; Appenzeller, J. Two-dimensional transistors with reconfigurable polarities for secure circuits. *Nat. Electron.* **2020**, *4*, 45–53. [CrossRef]
13. Chen, P.; Atallah, T.L.; Lin, Z.; Wang, P.; Lee, S.J.; Xu, J.; Huang, Z.; Duan, X.; Ping, Y.; Huang, Y.; et al. Approaching the intrinsic exciton physics limit in two-dimensional semiconductor diodes. *Nature* **2021**, *599*, 404–410. [CrossRef]
14. Nassiri Nazif, K.; Daus, A.; Hong, J.; Lee, N.; Vaziri, S.; Kumar, A.; Nitta, F.; Chen, M.E.; Kananian, S.; Islam, R.; et al. High-specific-power flexible transition metal dichalcogenide solar cells. *Nat. Commun.* **2021**, *12*, 7034. [CrossRef]
15. Wu, F.; Tian, H.; Shen, Y.; Hou, Z.; Ren, J.; Gou, G.; Sun, Y.; Yang, Y.; Ren, T.L. Vertical MoS_2 transistors with sub-1-nm gate lengths. *Nature* **2022**, *603*, 259–264. [CrossRef]
16. Chen, M.L.; Sun, X.; Liu, H.; Wang, H.; Zhu, Q.; Wang, S.; Du, H.; Dong, B.; Zhang, J.; Sun, Y.; et al. A FinFET with one atomic layer channel. *Nat. Commun.* **2020**, *11*, 1205. [CrossRef]
17. Chou, B.J.; Chung, Y.Y.; Yun, W.S.; Hsu, C.F.; Li, M.Y.; Su, S.K.; Liew, S.L.; Hou, V.D.; Chen, C.W.; Kei, C.C.; et al. High-performance monolayer MoS_2 nanosheet GAA transistor. *Nanotechnology* **2024**, *35*, 125204. [CrossRef] [PubMed]
18. Jiang, J.; Parto, K.; Cao, W.; Banerjee, K. Ultimate Monolithic-3D Integration With 2D Materials: Rationale, Prospects, and Challenges. *IEEE J. Electron Devices Soc.* **2019**, *7*, 878–887. [CrossRef]
19. Guo, Y.; Li, J.; Zhan, X.; Wang, C.; Li, M.; Zhang, B.; Wang, Z.; Liu, Y.; Yang, K.; Wang, H.; et al. Van der Waals polarity-engineered 3D integration of 2D complementary logic. *Nature* **2024**, *630*, 346–352. [CrossRef]
20. Lu, D.; Chen, Y.; Lu, Z.; Ma, L.; Tao, Q.; Li, Z.; Kong, L.; Liu, L.; Yang, X.; Ding, S.; et al. Monolithic three-dimensional tier-by-tier integration via van der Waals lamination. *Nature* **2024**, *630*, 340–345. [CrossRef] [PubMed]
21. Asselberghs, I.; Smets, Q.; Schram, T.; Groven, B.; Verreck, D.; Afzalian, A.; Arutchelvan, G.; Gaur, A.; Cott, D.; Maurice, T.; et al. Wafer-scale integration of double gated WS_2-transistors in 300mm Si CMOS fab. In Proceedings of the 2020 IEEE International Electron Devices Meeting (IEDM), San Francisco, CA, USA, 12–18 December 2020; pp. 40.42.41–40.42.44.
22. Ahmed, Z.; Afzalian, A.; Schram, T.; Jang, D.; Verreck, D.; Smets, Q.; Schuddinck, P.; Chehab, B.; Sutar, S.; Arutchelvan, G.; et al. Introducing 2D-FETs in Device Scaling Roadmap using DTCO. In Proceedings of the 2020 IEEE International Electron Devices Meeting (IEDM), San Francisco, CA, USA, 12–18 December 2020; pp. 22.25.21–22.25.24.
23. Tong, L.; Peng, Z.; Lin, R.; Li, Z.; Wang, Y.; Huang, X.; Xue, K.H.; Xu, H.; Liu, F.; Xia, H.; et al. 2D materials-based homogeneous transistor-memory architecture for neuromorphic hardware. *Science* **2021**, *373*, 1353–1358. [CrossRef]
24. Resta, G.V.; Balaji, Y.; Lin, D.; Radu, I.P.; Catthoor, F.; Gaillardon, P.E.; De Micheli, G. Doping-Free Complementary Logic Gates Enabled by Two-Dimensional Polarity-Controllable Transistors. *ACS Nano* **2018**, *12*, 7039–7047. [CrossRef]
25. Larentis, S.; Fallahazad, B.; Movva, H.C.P.; Kim, K.; Rai, A.; Taniguchi, T.; Watanabe, K.; Banerjee, S.K.; Tutuc, E. Reconfigurable Complementary Monolayer $MoTe_2$ Field-Effect Transistors for Integrated Circuits. *ACS Nano* **2017**, *11*, 4832–4839. [CrossRef]
26. Meng, Y.; Wang, W.; Wang, W.; Li, B.; Zhang, Y.; Ho, J. Anti-Ambipolar Heterojunctions: Materials, Devices, and Circuits. *Adv. Mater.* **2024**, *36*, e2306290. [CrossRef] [PubMed]
27. Fei, W.; Trommer, J.; Lemme, M.C.; Mikolajick, T.; Heinzig, A. Emerging reconfigurable electronic devices based on two-dimensional materials: A review. *InfoMat* **2022**, *4*, 12355. [CrossRef]
28. Zhao, Y.; Sun, H.; Sheng, Z.; Zhang, D.W.; Zhou, P.; Zhang, Z. Recent progress on ambipolar 2D semiconductors in emergent reconfigurable electronics and optoelectronics. *Chin. Phys. B* **2023**, *32*, 128505. [CrossRef]
29. Pan, C.; Wang, C.-Y.; Liang, S.-J.; Wang, Y.; Cao, T.; Wang, P.; Wang, C.; Wang, S.; Cheng, B.; Gao, A.; et al. Reconfigurable logic and neuromorphic circuits based on electrically tunable two-dimensional homojunctions. *Nat. Electron.* **2020**, *3*, 383–390. [CrossRef]
30. Lee, M.; Park, C.Y.; Hwang, D.K.; Kim, M.-g.; Lee, Y.T. Longitudinal and latitudinal split-gate field-effect transistors for NAND and NOR logic circuit applications. *npj 2D Mater. Appl.* **2022**, *6*, 45. [CrossRef]
31. Pan, J.; Wu, F.; Wang, Z.; Liu, S.; Guo, P.; Yin, J.; Zhao, B.; Tian, H.; Yang, Y.; Ren, T.L. Multibarrier Collaborative Modulation Devices with Ultra-High Logic Operation Density. *ACS Nano* **2024**, *18*, 28189–28197. [CrossRef]
32. Tsai, M.-Y.; Huang, C.-T.; Lin, C.-Y.; Lee, M.-P.; Yang, F.-S.; Li, M.; Chang, Y.-M.; Watanabe, K.; Taniguchi, T.; Ho, C.-H.; et al. A reconfigurable transistor and memory based on a two-dimensional heterostructure and photoinduced trapping. *Nat. Electron.* **2023**, *6*, 755–764. [CrossRef]

33. Leong, J.F.; Fang, Z.; Sivan, M.; Pan, J.; Tang, B.; Zamburg, E.; Thean, A.V.Y. N-P Reconfigurable Dual-Mode Memtransistors for Compact Bio-Inspired Feature Extractor with Inhibitory-Excitatory Spiking Capability. *Adv. Funct. Mater.* **2023**, *33*, 2302949. [CrossRef]
34. Mennel, L.; Symonowicz, J.; Wachter, S.; Polyushkin, D.K.; Molina-Mendoza, A.J.; Mueller, T. Ultrafast machine vision with 2D material neural network image sensors. *Nature* **2020**, *579*, 62–66. [CrossRef] [PubMed]
35. Wu, G.; Zhang, X.; Feng, G.; Wang, J.; Zhou, K.; Zeng, J.; Dong, D.; Zhu, F.; Yang, C.; Zhao, X.; et al. Ferroelectric-defined reconfigurable homojunctions for in-memory sensing and computing. *Nat. Mater.* **2023**, *22*, 1499–1506. [CrossRef] [PubMed]
36. Radisavljevic, B.; Radenovic, A.; Brivio, J.; Giacometti, V.; Kis, A. Single-layer MoS_2 transistors. *Nat. Nanotechnol.* **2011**, *6*, 147–150. [CrossRef] [PubMed]
37. Li, L.; Yu, Y.; Ye, G.J.; Ge, Q.; Ou, X.; Wu, H.; Feng, D.; Chen, X.H.; Zhang, Y. Black phosphorus field-effect transistors. *Nat. Nanotechnol.* **2014**, *9*, 372–377. [CrossRef]
38. Pal, A.; Chavan, T.; Jabbour, J.; Cao, W.; Banerjee, K. Three-dimensional transistors with two-dimensional semiconductors for future CMOS scaling. *Nat. Electron.* **2024**, *7*, 1147–1157. [CrossRef]
39. Wu, F.; Ren, J.; Yang, Y.; Yan, Z.; Tian, H.; Gou, G.; Wang, X.; Zhang, Z.; Yang, X.; Wu, X.; et al. A 10 nm Short Channel MoS_2 Transistor without the Resolution Requirement of Photolithography. *Adv. Electron. Mater.* **2021**, *7*, 2100543. [CrossRef]
40. Li, N.; Wang, Q.; Shen, C.; Wei, Z.; Yu, H.; Zhao, J.; Lu, X.; Wang, G.; He, C.; Xie, L.; et al. Large-scale flexible and transparent electronics based on monolayer molybdenum disulfide field-effect transistors. *Nat. Electron.* **2020**, *3*, 711–717. [CrossRef]
41. Tang, J.; Wang, Q.; Tian, J.; Li, X.; Li, N.; Peng, Y.; Li, X.; Zhao, Y.; He, C.; Wu, S.; et al. Low power flexible monolayer MoS_2 integrated circuits. *Nat. Commun.* **2023**, *14*, 3633. [CrossRef]
42. Lopez-Sanchez, O.; Lembke, D.; Kayci, M.; Radenovic, A.; Kis, A. Ultrasensitive photodetectors based on monolayer MoS_2. *Nat. Nanotechnol.* **2013**, *8*, 497–501. [CrossRef]
43. Wu, G.; Wang, X.; Chen, Y.; Wu, S.; Wu, B.; Jiang, Y.; Shen, H.; Lin, T.; Liu, Q.; Wang, X.; et al. $MoTe_2$ p-n Homojunctions Defined by Ferroelectric Polarization. *Adv. Mater.* **2020**, *32*, e1907937. [CrossRef] [PubMed]
44. Wu, G.; Abid, M.; Zerara, M.; Cho, J.; Choi, M.; Coileáin, C.Ó.; Hung, K.M.; Chang, C.R.; Shvets, I.V.; Wu, H.C. Miniaturized spectrometer with intrinsic long-term image memory. *Nat. Commun.* **2024**, *15*, 676. [CrossRef] [PubMed]
45. Zeng, S.; Liu, C.; Zhou, P. Transistor engineering based on 2D materials in the post-silicon era. *Nat. Rev. Electr. Eng.* **2024**, *1*, 335–348. [CrossRef]
46. Novoselov, K.S.; Geim, A.K.; Morozov, S.V.; Jiang, D.; Zhang, Y.; Dubonos, S.V.; Grigorieva, I.V.; Firsov, A.A. Electric field effect in atomically thin carbon films. *Science* **2004**, *306*, 666–669. [CrossRef] [PubMed]
47. Li, Y.; Kuang, G.; Jiao, Z.; Yao, L.; Duan, R. Recent progress on the mechanical exfoliation of 2D transition metal dichalcogenides. *Mater. Res. Express* **2022**, *9*, 122001. [CrossRef]
48. Lanza, M.; Smets, Q.; Huyghebaert, C.; Li, L.J. Yield, variability, reliability, and stability of two-dimensional materials based solid-state electronic devices. *Nat. Commun.* **2020**, *11*, 5689. [CrossRef]
49. Zhao, T.; Guo, J.; Li, T.; Wang, Z.; Peng, M.; Zhong, F.; Chen, Y.; Yu, Y.; Xu, T.; Xie, R.; et al. Substrate engineering for wafer-scale two-dimensional material growth: Strategies, mechanisms, and perspectives. *Chem. Soc. Rev.* **2023**, *52*, 1650–1671. [CrossRef]
50. Sheng, C.; Dong, X.; Zhu, Y.; Wang, X.; Chen, X.; Xia, Y.; Xu, Z.; Zhou, P.; Wan, J.; Bao, W. Two-Dimensional Semiconductors: From Device Processing to Circuit Integration. *Adv. Funct. Mater.* **2023**, *33*, 2304778. [CrossRef]
51. Briggs, N.; Subramanian, S.; Lin, Z.; Li, X.; Zhang, X.; Zhang, K.; Xiao, K.; Geohegan, D.; Wallace, R.; Chen, L.-Q.; et al. A roadmap for electronic grade 2D materials. *2D Mater.* **2019**, *6*, 022001. [CrossRef]
52. Zhu, J.; Park, J.H.; Vitale, S.A.; Ge, W.; Jung, G.S.; Wang, J.; Mohamed, M.; Zhang, T.; Ashok, M.; Xue, M.; et al. Low-thermal-budget synthesis of monolayer molybdenum disulfide for silicon back-end-of-line integration on a 200 mm platform. *Nat. Nanotechnol.* **2023**, *18*, 456–463. [CrossRef] [PubMed]
53. He, Y.; Andrade, A.F.; Menard-Moyon, C.; Bianco, A. Biocompatible 2D Materials via Liquid Phase Exfoliation. *Adv. Mater.* **2024**, *36*, e2310999. [CrossRef]
54. Liu, C.; Liu, T.; Zhang, Z.; Sun, Z.; Zhang, G.; Wang, E.; Liu, K. Understanding epitaxial growth of two-dimensional materials and their homostructures. *Nat. Nanotechnol.* **2024**, *19*, 907–918. [CrossRef]
55. Li, W.; Zhang, Y.; Long, X.; Cao, J.; Xin, X.; Guan, X.; Peng, J.; Zheng, X. Gas Sensors Based on Mechanically Exfoliated MoS_2 Nanosheets for Room-Temperature NO_2 Detection. *Sensors* **2019**, *19*, 2123. [CrossRef]
56. Park, J.H.; Lu, A.Y.; Shen, P.C.; Shin, B.G.; Wang, H.; Mao, N.; Xu, R.; Jung, S.J.; Ham, D.; Kern, K.; et al. Synthesis of High-Performance Monolayer Molybdenum Disulfide at Low Temperature. *Small Methods* **2021**, *5*, e2000720. [CrossRef] [PubMed]
57. Chen, H.; Xue, X.; Liu, C.; Fang, J.; Wang, Z.; Wang, J.; Zhang, D.W.; Hu, W.; Zhou, P. Logic gates based on neuristors made from two-dimensional materials. *Nat. Electron.* **2021**, *4*, 399–404. [CrossRef]
58. Das, S.; Appenzeller, J. WSe_2 field effect transistors with enhanced ambipolar characteristics. *Appl. Phys. Lett.* **2013**, *103*, 103501. [CrossRef]

59. Lin, Y.F.; Xu, Y.; Wang, S.T.; Li, S.L.; Yamamoto, M.; Aparecido-Ferreira, A.; Li, W.; Sun, H.; Nakaharai, S.; Jian, W.B.; et al. Ambipolar $MoTe_2$ transistors and their applications in logic circuits. *Adv. Mater.* **2014**, *26*, 3263–3269. [CrossRef]
60. Zhu, W.; Yogeesh, M.N.; Yang, S.; Aldave, S.H.; Kim, J.S.; Sonde, S.; Tao, L.; Lu, N.; Akinwande, D. Flexible black phosphorus ambipolar transistors, circuits and AM demodulator. *Nano Lett.* **2015**, *15*, 1883–1890. [CrossRef] [PubMed]
61. Du, Y.; Liu, H.; Deng, Y.; Ye, P.D. Device perspective for black phosphorus field-effect transistors: Contact resistance, ambipolar behavior, and scaling. *ACS Nano* **2014**, *8*, 10035–10042. [CrossRef]
62. Chen, L.; Li, S.; Feng, X.; Wang, L.; Huang, X.; Tee, B.C.K.; Ang, K.W. Gigahertz Integrated Circuits Based on Complementary Black Phosphorus Transistors. *Adv. Electron. Mater.* **2018**, *4*, 1800274. [CrossRef]
63. Tian, H.; Li, Y.-x.; Li, L.; Wang, X.; Liang, R.; Yang, Y.; Ren, T.-L. Negative Capacitance Black Phosphorus Transistors With Low SS. *IEEE Trans. Electron Dev.* **2019**, *66*, 1579–1583. [CrossRef]
64. Chen, X.; Xie, Y.; Sheng, Y.; Tang, H.; Wang, Z.; Wang, Y.; Wang, Y.; Liao, F.; Ma, J.; Guo, X.; et al. Wafer-scale functional circuits based on two dimensional semiconductors with fabrication optimized by machine learning. *Nat. Commun.* **2021**, *12*, 5953. [CrossRef] [PubMed]
65. Fan, D.; Li, W.; Qiu, H.; Xu, Y.; Gao, S.; Liu, L.; Li, T.; Huang, F.; Mao, Y.; Zhou, W.; et al. Two-dimensional semiconductor integrated circuits operating at gigahertz frequencies. *Nat. Electron.* **2023**, *6*, 879–887. [CrossRef]
66. Wang, Y.; Kim, J.C.; Li, Y.; Ma, K.Y.; Hong, S.; Kim, M.; Shin, H.S.; Jeong, H.Y.; Chhowalla, M. P-type electrical contacts for 2D transition-metal dichalcogenides. *Nature* **2022**, *610*, 61–66. [CrossRef]
67. Pendurthi, R.; Sakib, N.U.; Sadaf, M.U.K.; Zhang, Z.; Sun, Y.; Chen, C.; Jayachandran, D.; Oberoi, A.; Ghosh, S.; Kumari, S.; et al. Monolithic three-dimensional integration of complementary two-dimensional field-effect transistors. *Nat. Nanotechnol.* **2024**, *19*, 970–977. [CrossRef] [PubMed]
68. Wang, F.; Pei, K.; Li, Y.; Li, H.; Zhai, T. 2D Homojunctions for Electronics and Optoelectronics. *Adv. Mater.* **2021**, *33*, e2005303. [CrossRef] [PubMed]
69. Li, X.; Zhou, P.; Hu, X.; Rivers, E.; Watanabe, K.; Taniguchi, T.; Akinwande, D.; Friedman, J.S.; Incorvia, J.A.C. Cascaded Logic Gates Based on High-Performance Ambipolar Dual-Gate WSe_2 Thin Film Transistors. *ACS Nano* **2023**, *17*, 12798–12808. [CrossRef] [PubMed]
70. Zhou, Y.; Fu, J.; Chen, Z.; Zhuge, F.; Wang, Y.; Yan, J.; Ma, S.; Xu, L.; Yuan, H.; Chan, M.; et al. Computational event-driven vision sensors for in-sensor spiking neural networks. *Nature Electronics* **2023**, *6*, 870–878. [CrossRef]
71. Bu, T.; Duan, X.; Liu, C.; Su, W.; Hong, X.; Hong, R.; Zhou, X.; Liu, Y.; Fan, Z.; Zou, X.; et al. Electrically Dynamic Configurable WSe_2 Transistor and the Applications in Photodetector. *Adv. Funct. Mater.* **2023**, *33*, 2305490. [CrossRef]
72. Xu, K.; Zhao, Y.; Lin, Z.; Long, Y.; Wang, Y.; Chan, M.; Chai, Y. Doping of two-dimensional MoS_2 by high energy ion implantation. *Semicond. Sci. Technol.* **2017**, *32*, 124002. [CrossRef]
73. Nipane, A.; Karmakar, D.; Kaushik, N.; Karande, S.; Lodha, S. Few-Layer MoS_2 p-Type Devices Enabled by Selective Doping Using Low Energy Phosphorus Implantation. *ACS Nano* **2016**, *10*, 2128–2137. [CrossRef]
74. Peng, R.; Wu, Y.; Wang, B.; Shi, R.; Xu, L.; Pan, T.; Guo, J.; Zhao, B.; Song, C.; Fan, Z.; et al. Programmable graded doping for reconfigurable molybdenum ditelluride devices. *Nat. Electron.* **2023**, *6*, 852–861. [CrossRef]
75. Jin, T.; Gao, J.; Wang, Y.; Zheng, Y.; Sun, S.; Liu, L.; Lin, M.; Chen, W. Two-dimensional reconfigurable electronics enabled by asymmetric floating gate. *Nano Res.* **2022**, *15*, 4439–4447. [CrossRef]
76. Wu, F.; Tian, H.; Yan, Z.; Ren, J.; Hirtz, T.; Gou, G.; Shen, Y.; Yang, Y.; Ren, T.L. Gate-Tunable Negative Differential Resistance Behaviors in a hBN-Encapsulated BP-MoS_2 Heterojunction. *ACS Appl. Mater. Interfaces* **2021**, *13*, 26161–26169. [CrossRef]
77. Shim, J.; Oh, S.; Kang, D.H.; Jo, S.H.; Ali, M.H.; Choi, W.Y.; Heo, K.; Jeon, J.; Lee, S.; Kim, M.; et al. Phosphorene/rhenium disulfide heterojunction-based negative differential resistance device for multi-valued logic. *Nat. Commun.* **2016**, *7*, 13413. [CrossRef] [PubMed]
78. Nourbakhsh, A.; Zubair, A.; Dresselhaus, M.S.; Palacios, T. Transport Properties of a MoS_2/WSe_2 Heterojunction Transistor and Its Potential for Application. *Nano Lett.* **2016**, *16*, 1359–1366. [CrossRef]
79. Huang, M.; Li, S.; Zhang, Z.; Xiong, X.; Li, X.; Wu, Y. Multifunctional high-performance van der Waals heterostructures. *Nat. Nanotechnol.* **2017**, *12*, 1148–1154. [CrossRef]
80. Seo, S.; Cho, J.I.; Jung, K.S.; Andreev, M.; Lee, J.H.; Ahn, H.; Jung, S.; Lee, T.; Kim, B.; Lee, S.; et al. A Van Der Waals Reconfigurable Multi-Valued Logic Device and Circuit Based on Tunable Negative-Differential-Resistance Phenomenon. *Adv. Mater.* **2022**, *34*, e2202799. [CrossRef]
81. Li, Y.; Wang, Y.; Huang, L.; Wang, X.; Li, X.; Deng, H.X.; Wei, Z.; Li, J. Anti-Ambipolar Field-Effect Transistors Based On Few-Layer 2D Transition Metal Dichalcogenides. *ACS Appl. Mater. Interfaces* **2016**, *8*, 15574–15581. [CrossRef]
82. Shingaya, Y.; Zulkefli, A.; Iwasaki, T.; Hayakawa, R.; Nakaharai, S.; Watanabe, K.; Taniguchi, T.; Wakayama, Y. Dual-Gate Anti-Ambipolar Transistor with Van der Waals ReS_2/WSe_2 Heterojunction for Reconfigurable Logic Operations. *Adv. Electron. Mater.* **2022**, *9*, 2200704. [CrossRef]

83. Cavallini, M.; Gentili, D. Atomic Vacancies in Transition Metal Dichalcogenides: Properties, Fabrication, and Limits. *Chempluschem* **2022**, *87*, e202100562. [CrossRef] [PubMed]
84. Zhou, W.; Zou, X.; Najmaei, S.; Liu, Z.; Shi, Y.; Kong, J.; Lou, J.; Ajayan, P.M.; Yakobson, B.I.; Idrobo, J.C. Intrinsic structural defects in monolayer molybdenum disulfide. *Nano Lett* **2013**, *13*, 2615–2622. [CrossRef] [PubMed]
85. Mahjouri-Samani, M.; Liang, L.; Oyedele, A.; Kim, Y.S.; Tian, M.; Cross, N.; Wang, K.; Lin, M.W.; Boulesbaa, A.; Rouleau, C.M.; et al. Tailoring Vacancies Far Beyond Intrinsic Levels Changes the Carrier Type and Optical Response in Monolayer $MoSe_{2-x}$ Crystals. *Nano Lett* **2016**, *16*, 5213–5220. [CrossRef] [PubMed]
86. Wu, X.; Ge, R.; Akinwande, D.; Lee, J.C. Understanding of multiple resistance states by current sweeping in MoS_2-based non-volatile memory devices. *Nanotechnology* **2020**, *31*, 465206. [CrossRef] [PubMed]
87. Peng, B.; Ning, Z.; Zhang, H.; Shao, H.; Xu, Y.; Ni, G.; Zhu, H. Beyond Perturbation: Role of Vacancy-Induced Localized Phonon States in Thermal Transport of Monolayer MoS_2. *J. Phys. Chem. C* **2016**, *120*, 29324–29331. [CrossRef]
88. Jo, S.B.; Kang, J.; Cho, J.H. Recent Advances on Multivalued Logic Gates: A Materials Perspective. *Adv. Sci. (Weinh.)* **2021**, *8*, 2004216. [CrossRef] [PubMed]
89. Seo, S.; Koo, J.; Choi, J.-W.; Heo, K.; Andreev, M.; Lee, J.-J.; Lee, J.-H.; Cho, J.-I.; Kim, H.; Yoo, G.; et al. Controllable potential barrier for multiple negative-differential-transconductance and its application to multi-valued logic computing. *npj 2D Mater. Appl.* **2021**, *5*, 32. [CrossRef]
90. Yi, J.; Sun, X.; Zhu, C.; Li, S.; Liu, Y.; Zhu, X.; You, W.; Liang, D.; Shuai, Q.; Wu, Y.; et al. Double-Gate MoS_2 Field-Effect Transistors with Full-Range Tunable Threshold Voltage for Multifunctional Logic Circuits. *Adv. Mater.* **2021**, *33*, e2101036. [CrossRef]
91. Sun, X.; Zhu, C.; Yi, J.; Xiang, L.; Ma, C.; Liu, H.; Zheng, B.; Liu, Y.; You, W.; Zhang, W.; et al. Reconfigurable logic-in-memory architectures based on a two-dimensional van der Waals heterostructure device. *Nat. Electron.* **2022**, *5*, 752–760. [CrossRef]
92. Zeng, D.; Ding, R.; Liu, G.; Lu, H.; Zhang, M.; Xue, Z.; Tian, Z.; Di, Z. Side-Gate BN-MoS_2 Transistor for Reconfigurable Multifunctional Electronics. *Adv. Electron. Mater.* **2023**, *10*, 2300621. [CrossRef]
93. Wang, H.; Bao, L.; Guzman, R.; Wu, K.; Wang, A.; Liu, L.; Wu, L.; Chen, J.; Huan, Q.; Zhou, W.; et al. Ultrafast-Programmable 2D Homojunctions Based on van der Waals Heterostructures on a Silicon Substrate. *Adv. Mater.* **2023**, *35*, e2301067. [CrossRef]
94. Ma, Z.; Yuan, P.; Li, L.; Tang, X.; Li, X.; Zhang, S.; Yu, L.; Jiang, Y.; Song, X.; Xia, C. Optoelectronic Reconfigurable Logic Gates Based on Two-Dimensional Vertical Field-Effect Transistors. *Nano Lett.* **2024**, *24*, 14058–14065. [CrossRef]
95. Bach, T.P.A.; Cho, S.; Kim, H.; Nguyen, D.A.; Im, H. 2D van der Waals Heterostructure with Tellurene Floating-Gate for Wide Range and Multi-Bit Optoelectronic Memory. *ACS Nano* **2024**, *18*, 4131–4139. [CrossRef]
96. Sahu, M.C.; Sahoo, S.; Mallik, S.K.; Jena, A.K.; Sahoo, S. Multifunctional 2D MoS_2 Optoelectronic Artificial Synapse with Integrated Arithmetic and Reconfigurable Logic Operations for In-Memory Neuromorphic Computing Applications. *Adv. Mater. Technol.* **2022**, *8*, 2201125. [CrossRef]
97. Yoo, C.; Ko, T.-J.; Kaium, M.G.; Martinez, R.; Islam, M.M.; Li, H.; Kim, J.H.; Cao, J.; Acharya, M.; Roy, T.; et al. A minireview on 2D materials-enabled optoelectronic artificial synaptic devices. *APL Mater.* **2022**, *10*, 070702. [CrossRef]
98. Mukherjee, S.; Dutta, D.; Ghosh, A.; Koren, E. Graphene-In_2Se_3 van der Waals Heterojunction Neuristor for Optical In-Memory Bimodal Operation. *ACS Nano* **2023**, *17*, 22287–22298. [CrossRef] [PubMed]
99. Hu, J.; Li, H.; Zhang, Y.; Zhou, J.; Zhao, Y.; Xu, Y.; Yu, B. Reconfigurable Neuromorphic Computing with 2D Material Heterostructures for Versatile Neural Information Processing. *Nano Lett.* **2024**, *24*, 9391–9398. [CrossRef]
100. Yao, C.; Wu, G.; Huang, M.; Wang, W.; Zhang, C.; Wu, J.; Liu, H.; Zheng, B.; Yi, J.; Zhu, C.; et al. Reconfigurable Artificial Synapse Based on Ambipolar Floating Gate Memory. *ACS Appl. Mater. Interfaces* **2023**, *15*, 23573–23582. [CrossRef] [PubMed]
101. Wang, C.Y.; Liang, S.J.; Wang, S.; Wang, P.; Li, Z.; Wang, Z.; Gao, A.; Pan, C.; Liu, C.; Liu, J.; et al. Gate-tunable van der Waals heterostructure for reconfigurable neural network vision sensor. *Sci. Adv.* **2020**, *6*, eaba6173. [CrossRef]
102. Zhang, W.; Gao, B.; Tang, J.; Yao, P.; Yu, S.; Chang, M.-F.; Yoo, H.-J.; Qian, H.; Wu, H. Neuro-inspired computing chips. *Nat. Electron.* **2020**, *3*, 371–382. [CrossRef]
103. Dang, Z.; Guo, F.; Wang, Z.; Jie, W.; Jin, K.; Chai, Y.; Hao, J. Object Motion Detection Enabled by Reconfigurable Neuromorphic Vision Sensor under Ferroelectric Modulation. *ACS Nano* **2024**, *18*, 27727–27737. [CrossRef]
104. Zhang, Z.; Wang, S.; Liu, C.; Xie, R.; Hu, W.; Zhou, P. All-in-one two-dimensional retinomorphic hardware device for motion detection and recognition. *Nat. Nanotechnol.* **2022**, *17*, 27–32. [CrossRef] [PubMed]

MDPI AG
Grosspeteranlage 5
4052 Basel
Switzerland
Tel.: +41 61 683 77 34

Nanomaterials Editorial Office
E-mail: nanomaterials@mdpi.com
www.mdpi.com/journal/nanomaterials

www.ingramcontent.com/pod-product-compliance
Lightning Source LLC
Chambersburg PA
CBHW062045220926
51850CB00010B/2663
* 9 7 8 3 7 2 5 8 6 2 8 6 3 *